The Periodic Table of the Elements of Green and Sustainable Chemistry

Copyright © 2019

by

Paul T. Anastas and Julie B. Zimmerman

ISBN: 978-1-7345463-0-9

First Edition: January 2020

Published by Press Zero, Madison, CT, USA, 06437

Acknowledgements

The authors wish to thank the entirety of the international green chemistry community for their efforts in creating a sustainable tomorrow.

The authors would also like to thank Dr. Evan Beach for his thoughtful and constructive contributions during the editing of this volume, Ms. Kimberly Chapman for her work on the graphics for the table, and Mr. John Pfannkuchen for his work on the layout of the book. In addition, the authors would like to thank the Royal Society of Chemistry for their continued support for the field of green chemistry.

To Kennedy and Aquinnah

CONTENTS

PREFACE

The field of Green Chemistry emerged in the early 1990s as an approach to using the power and potential of chemistry to design the next generation of products and processes such that they are good for humans, the environment, society, and the economy. Its innovations are felt around the world in improving the products of our daily lives, the way we make our medicines and grow our food, and how we generate and store our energy. In recent years, sustainable chemistry has been introduced to describe those important aspects beyond the science that are necessary for green chemistry to make a positive impact on the world at the scale and timeframe necessary to advance sustainability.

In celebration of the 150th anniversary of the Periodic Table of the Elements, a new metaphorical construct has been assembled to illustrate the "elements" that will be crucial on the path to a sustainable future. This is the basis of "The Periodic Table of the Elements of Green and Sustainable Chemistry."

It's 150 years since Mendeleev introduced the Periodic Table of the Elements. the periodic table allowed the prediction how things interact; how they behave. The periodic table, acts as an organizing framework for how we see how things will react with one another and they are fundamental properties. And all of us recognize that chemistry is everywhere in ubiquitous. "Chemistry is everywhere and it's everything," is the common phrase amongst many chemists. It's everything we see touch and feel. It's everything we use the basis of our society and our economy. Many refer to chemistry it as the central science; the central and essential science.

The Periodic Table of the Elements of

Humanitarian Elements

1 A Appropriate Technologies for the Developing World		
3 Cw Chemistry for Wellness	**4 Dd** Design to Avoid Dependency	
11 Sw Access to Safe and Reliable Water	**12 Fg** Ensure Access to Material Resources for Future Generations	

Green Chemistry Green Engineering Elements

- ■ Prevent Waste
- ■ Atom Economy
- ■ Less Hazardous Synthesis
- ■ Molecular Design
- ■ Solvents/Aux
- ■ Energy
- ■ Renewable Feed Stocks
- ■ Catalysis
- ■ Degradation
- ■ Measurement and Awareness

19 **Bf** Chemistry for Benign Food Production and Nutrition	20 **Tc** Transparency for Chemical Communication	21 **Wu** Waste Material Utilization and Valorization	22 **Sa** Molecular Self-Assembly	23 **Ru** Reduce use of Hazardous Materials	24 **Dg** Design Guidelines	25 **Aq** Aqueous and Biobased Solvents	26 **Ee** Energy and Material Efficient Synthesis and Processing	27 **Ib** Integrated Biorefinery
37 **J** Ensure Environmental Justice, Security, and Equitable Opportunities	38 **Cs** Chemistry for Sustainable Building and Buildings	39 **Op** One-Pot Synthesis	40 **Ip** Integrated Processes	41 **Gc** In-Situ Generation & Consumption of Hazardous Materials	42 **Cm** Computational Models	43 **Il** Ionic Liquids / Non-Volatile Solvents	44 **R** Renewable / Carbon-Free Energy Inputs	45 **C** Carbon Dioxide and other C1 Feedstocks
55 **Pc** Chemistry to Preserve Natural Carbon and Other Biogeochemical Cycles	56 **Ic** An Individual's Molecular Code Belongs to that Individual	57 **Pi** Process Intensification	58 **As** Additive Synthesis	59 **Ch** C-H Bond Functionalization	60 **Ba** Bioavailability / ADME	61 **Sc** Sub- and Super-Critical Fluids	62 **Es** Energy Storage / Transmission Materials	63 **Sb** Synthetic Biology
73 **Wo** No Chemicals of War or Oppression	74 **Nc** Molecular Codes of Nature Belong to the World	75 **Ss** Self-Separation	76 **W** Non-Covalent Derivatives / Weak Force Transformation	77 **Is** Inherent Safety and Security	78 **Ts** High Throughput Screening (Empirical / In Vivo / In Vitro)	79 **S** "Smart" Solvents (Obedient, Tunable)	80 **V** Waste Energy Utilization and Valorization	81 **Bt** Biologically-Enabled Transformation

GREEN AND SUSTAINABLE CHEMISTRY

ENABLING SYSTEM CONDITIONS

NOBLE ELEMENTS

Legend:
- Conceptual Frameworks
- Economic and Market Forces
- Metrics
- Policies and Regulations
- Tools

					2 **Ho** Hippocratic Oath for Chemistry
5 **B** Biomimicry	6 **Cb** Life Cycle Cost-Benefit Analysis	7 **Ae** Atom Economy	8 **Pr** Extended Producer Responsibility	9 **Ea** Epidemiological Analysis and Ecosystem Health	10 **P** Design for Posterity
13 **Ce** Circular Economy	14 **Fc** Full Cost Accounting	15 **Ef** E-Factor	16 **Pb** Property Based Regulation	17 **Aa** Alternatives Assessment	18 **Lp** Life-Compatible Products & Processes

28 **E** Enzymes	29 **Bm** Benign Metabolites	30 **Sn** Sensors	31 **Bd** Benign by Design	32 **Hc** Harm Charge / Carbon Tax	33 **Ff** F-Factor	34 **Ct** Chemical Transparency	35 **Lc** Life Cycle Assessment	36 **Z** Zero Waste
46 **Ac** Earth Abundant Metal Catalysis	47 **Md** Molecular Degradation Triggers	66 **Ex** Exposome	49 **Ie** Industrial Ecology	50 **Dc** Depletion Charge	51 **Ql** Qualitative Metrics	52 **Cl** Chemical Leasing	53 **So** Solvent Selection Screens	54 **Fi** Chemistry is Equitable and Fully Inclusive
64 **Ht** Heterogeneous Catalysis	65 **Dp** Degradable Polymers and Other Materials	48 **Co** In-Process Control and Optimization	67 **Tg** Trans-Generational Design	68 **Rf** Sustained Research Funding	69 **Qn** Quantitative Metrics	70 **Se** Self-Enforcing Regulations	71 **Cf** Chemical Footprinting	72 **De** Benefits Distributed Equitably
82 **Hm** Homogeneous Catalysis	83 **Pd** Prediction and Design Tools	84 **Ga** Green Analytical Chemistry	85 **Be** Bio-Based Economy	86 **Ci** Capital Investment	87 **Bb** Chemical Body Burden	88 **I** Innovation Ecosystem Translation from Lab to Commerce	89 **Et** Education in Toxicology and Systems Thinking	90 **K** Extraordinary Chemical Knowledge Comes with Extraordinary Responsibility

Anastas, P. T.; Zimmerman, J. B. The Periodic Table of the Elements of Green and Sustainable Chemistry, *Green Chemistry* **2019**, *21*, 6545-6566. DOI: 10.1039/C9GC01293A

x

Introduction

When thinking about the brilliance of how creative chemistry has transformed civilizations in terms of volume and efficiency of food production, access to energy, invention of new medicines, ability to travel far distances, and the insights gained through its role in the information revolution, one soon recognizes why chemistry is often referred to as "the central science". But certainly, it is not merely central to various industrial sectors or central to enabling a range of academic disciplines. It must be central to more than just that.

What it is central to is perhaps the most important question of our time. Because when recognizing the fundamental nature and the power of chemistry, we recognize that it is central to whether we will meet the greatest challenges of current and future generations. When considering the great urgent goals of sustainability, chemistry is central to the difference between success and failure, survival and extinction. It is that fundamental.

In recent decades, the field of green chemistry has emerged with all of its technological and commercial achievements, with all of the new perspectives that green chemistry brings, all of the new inventions and innovations, new reactions and transformations of forms of matter, and the motivation and impetus behind all of this activity becomes clear. We find the answer to the question: "Why do we do what we do?"

And at the foundation of everything we do are the humanitarian issues, the humanitarian goals, the human species and the ecosystems, biosphere, and geosphere upon which the human species relies.

Humanitarian Elements

The following elements of Green and Sustainable Chemistry emphasize core humanitarian aims and principles. Since chemistry has existed as a discipline, it has played a central role in meeting fundamental human needs. As we continue to press against the limits of natural systems, it will remain a challenge to ensure basic needs such as food, water, security, and shelter for future generations. Green and Sustainable Chemistry should also strive to ensure that risks and benefits are equally shared among populations.

1 **A** Appropriate Technologies for the Developing World	
3 **Cw** Chemistry for Wellness	**4** **Dd** Design to Avoid Dependency
11 **Sw** Access to Safe and Reliable Water	**12** **Fg** Ensure Access to Material Resources for Future Generations
19 **Bf** Chemistry for Benign Food Production and Nutrition	**20** **Tc** Transparency for Chemical Communication
37 **J** Ensure Environmental Justice, Security, and Equitable Opportunities	**38** **Cs** Chemistry for Sustainable Building and Buildings
55 **Pc** Chemistry to Preserve Natural Carbon and Other Biogeochemical Cycles	**56** **Ic** An Individual's Molecular Code Belongs to that Individual
73 **Wo** No Chemicals of War or Oppression	**74** **Nc** Molecular Codes of Nature Belong to the World

Appropriate Technologies for the Developing World

The invention and deployment of chemical technologies over the past two centuries has transformed modern life for a segment of the global population. However, many of the technologies developed in the past and still being developed today are not benefiting or improving the well-being of a large percentage of the world's population. This is due to factors that include particular resource flows, capital flows, infrastructure requirements, human capital, and more.

Developing appropriate technologies means understanding the context in which a technology will be deployed to ensure that it can bring benefit. Designing a technology to be viable in the industrialized world or the developing world, a major metropolis or a tribal village, requires systems thinking and thoughtful design. Designing technologies for the appropriate context is a challenge that needs to be met if benefit is to be equitable.

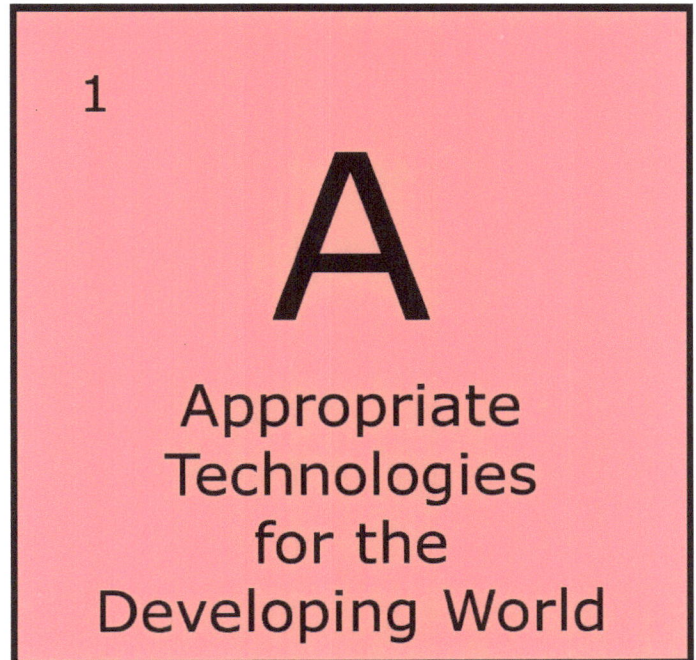

1

A

Appropriate Technologies for the Developing World

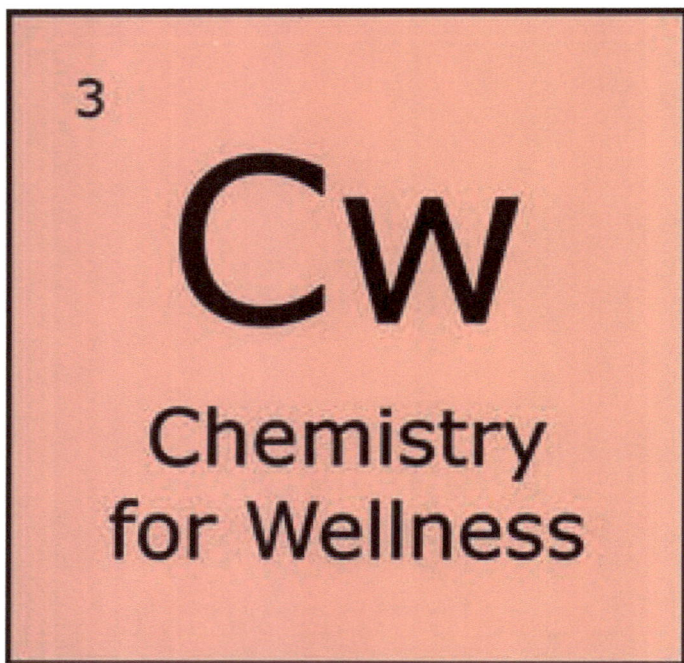

Chemistry for Wellness

Chemistry has been at the forefront of modern Western medicines with the vast majority of effort focused on treating disease. However, treating disease is not the same as preserving wellness. Chemistry has been at the forefront of increased food production. However, increased food supply and production of caloric content is not the same as the realization of access to nutritional health. Chemistry has been at the forefront of imaging, diagnostics, and sensors but measuring and monitoring a problem is not the same as solving it or preventing it. Chemistry can play a role in significant and timely efforts to rigorously understand and deliver nutraceuticals, probiotics, and other beneficial supplements. Chemists also need to work toward elucidating and enhancing positive mechanistic pathways toward proactive health.

3

Cw

Chemistry
for Wellness

Access to Safe and Reliable Water

One of the essential materials of life is water. Access to suitable water for drinking, sanitation, and hygiene is necessary for life, health, and well-being. Chemistry has been critical in historical approaches to water disinfection and improving water quality from removal of harmful contaminants to addressing odor and taste concerns. Chemistry will be required to realize the future methods, process, and materials that enable healthful drinking water without the unintended consequences of harmful disinfection by-products. Sanitation is an equally important imperative to avoid the spread of disease by exposure to pathogens. These services must be supplied in ways that are mindful of equity, safety, and dignity. The enabling infrastructure and processes for supply and treatment of potable and non-potable water will all depend on chemical innovations to ensure appropriate quantity and quality for drinking, sanitation, hygiene, or any other intended use.

11

Sw

Access to Safe and Reliable Water

Chemistry for Benign Food Production and Nutrition

19	
Bf	
Chemistry for Benign Food Production and Nutrition	

While the great increases in the efficiency of food production have been astounding and life-saving, there have been unintended consequences from the over-use of fertilizers and the use of pesticides and herbicides that harm beneficial insects and plants and damage ecosystems.

Chemical and non-chemical alternatives are being developed and need to be implemented at larger scale. These range from targeted bio-pesticides to herbicides that focus on specific pests and plants to pesticides designed to degrade rapidly into non-toxic degradation products. Integrated pest management systems have been shown to be efficacious and economically practical without the environmental damage.

Ensure Environmental Justice, Security, and Equitable Opportunities

Proximity to chemical manufacturing, processing, and use has historically come with disproportionate risk to the surrounding communities and ecosystems; this includes the exportation of banned chemicals and hazardous waste to communities with few other economic opportunities.

Individuals should have the right to live in homes and communities that are not placed at increased risk due to transportation or production of chemicals or inadequate protection of facilities. Use of highly hazardous chemicals increases vulnerability to accidents, natural disasters, or intentional acts of violence and terror. Adopting chemistries that are inherently benign is the most effective means of ensuring that communities are not placed at undue risk from the chemical enterprise. Case studies have shown that there are safer alternatives to toxic gases, VOCs, pesticides, and cleaning chemicals implicated in previous incidents.

37

J

Ensure Environmental Justice, Security, and Equitable Opportunities

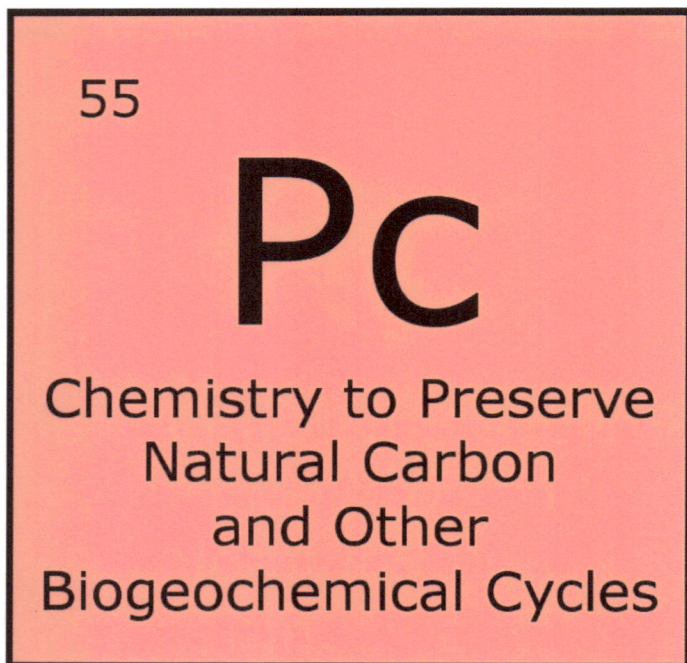

55

Pc

Chemistry to Preserve Natural Carbon and Other Biogeochemical Cycles

Chemistry to Preserve Natural Carbon and Other Biogeochemical Cycles

Integrated systems are the basis of the planet's natural biogeochemical cycles. From carbon to nitrogen to water and beyond, these chemical cycles and chemical processes are adversely impacted by the human interferences and interactions with these cycles. Human chemistry must be designed such that there will be no perturbation of the natural biogeochemical cycles and their interlinkages. Through thoughtful consideration of human material and energy utilization, we can protect the critical natural cycles essential to the preservation of life and all of its diversity on the planet.

No Chemicals of War or Oppression

The history of the intentional use of weaponized chemicals is recognized as uncivilized and the continuation of the universal rejection of this practice is essential to a civilized society. Likewise, the use of chemicals for carrying out the death penalty or for the oppression of individuals through chemical lobotomies or chemical castration must not happen. Designing, making, and utilizing chemicals for the purpose of mental or physical control over individuals is inconsistent with the peaceful uses of chemistry.

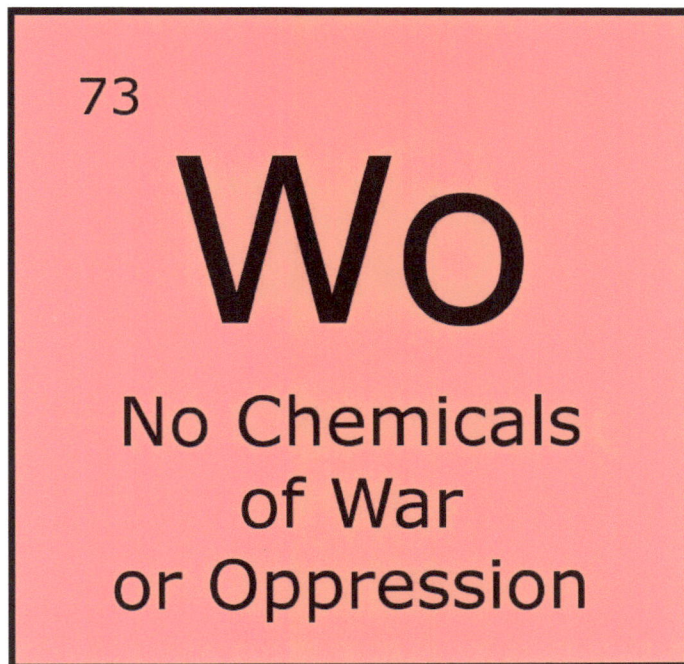

73

Wo

No Chemicals of War or Oppression

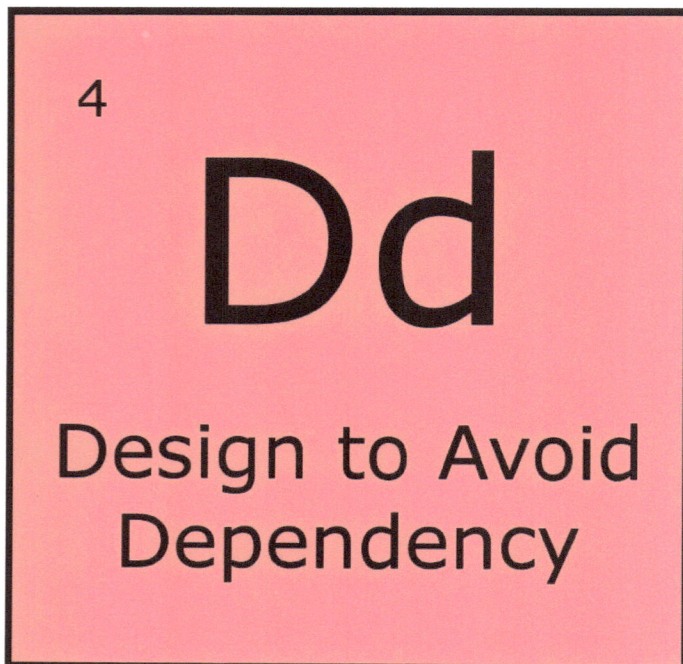

4

Dd
Design to Avoid Dependency

Design to Avoid Dependency

Some molecules can be addictive to humans. They can cause physical and neurological changes that result in dependency, and its subsequent impairment. Chemists have the duty to establish knowledge and awareness of the mechanisms of dependency in order to ensure that the molecules that are created do not result in human dependency and addiction.

Economic dependency is a separate yet important issue. Molecules should not be designed to be essential and inseparable to livelihood and wellness and embedded in systems that provide critical items like food, clean water, and medicine.

Chemical dependency is not limited to the human species; it can also impact other living things both animal and plant. Dependency needs to be understood such that design for variety and resilience is pursued preferably.

Ensure Access to Material Resources for Future Generations

The basic elements will largely not be created nor destroyed – with the exception of radioactive decay. However, the ability to access resources that are fundamental to the chemical and material infrastructure of our society and economy can be greatly impacted and diminished through irresponsible use. The combustion of valuable fossil fuels, the dissipation of phosphorus, and the diffusion of rare earth elements will make our resource base more difficult to access for future generations.

12

Fg

Ensure Access to Material Resources for Future Generations

11

20

Tc

Transparency
for Chemical
Communication

Transparency for Chemical Communication

The potential benefits and the potential harm that can be provided by chemistry are too immense and powerful to be cloaked in the darkness of jargon. Yet, as the evolution of chemistry proceeds, the greater the opacity of the science to the general public. The responsibility of molecular scientists – and scientists generally – is not merely to discover, invent, and understand, but also to effectively convey that understanding as clearly and transparently as possible. To be clear, this is not the same as "as clearly and transparently as convenient." The task of communication is not a secondary task but rather one of equal status in the scientific pursuit.

Chemistry for Sustainable Building and Buildings

The structures that are used to house people and their activities should be designed, constructed, and maintained to provide resilient protection while being conducive to the health and wellness of the occupants and the surrounding ecosystems. From the molecular basis of the materials throughout their life cycle to the systems that provide the requisite water, climate control, and lighting, chemistry has a significant role to play in efficiently providing safe indoor environments for humans and their activities while considering the surrounding landscape. Further, buildings can be designed to be modular, providing for future adaptions, deconstruction, and reuse to avoid end of life material waste.

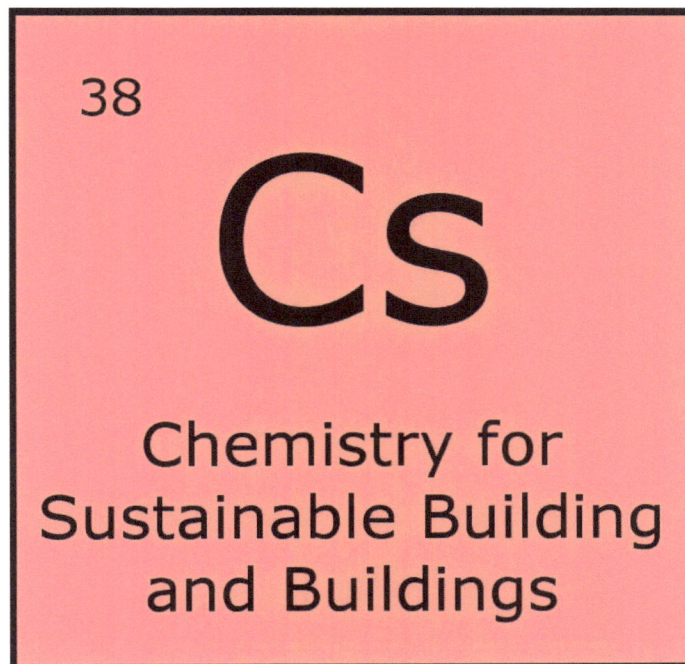

38

Cs

Chemistry for Sustainable Building and Buildings

An Individual's Molecular Code Belongs to that Individual

56	
Ic	
An Individual's Molecular Code Belongs to that Individual	

Every individual possesses the unique molecular code that is the underlying basis of their individuality and their identity. Further, this code has the power and potential for understanding health and wellness in a fundamental way, at the molecular scale. While this molecular code may contain information and functionality that can result in value, every individual is sovereign over their code and this sovereignty cannot be taken from them by others. As the ability to understand, manipulate, and utilize the various levels of this biological code increases, it will be ever more essential to guarantee that control of this code remains with the individual.

Molecular Codes of Nature Belong to the World

No human wrote any of the genetic codes of Nature and no human or group of humans can own it. Humans played no role in the billions of years of the design of genomes, bio/geo material structure, natural transformations and self-assembly. The inventions of Nature belong to Nature. The codes of Nature belong to Nature. The act of a human recognizing its brilliance and deeming it a discovery is not justification to claiming ownership and control or in any way limiting access to this brilliance by everyone.

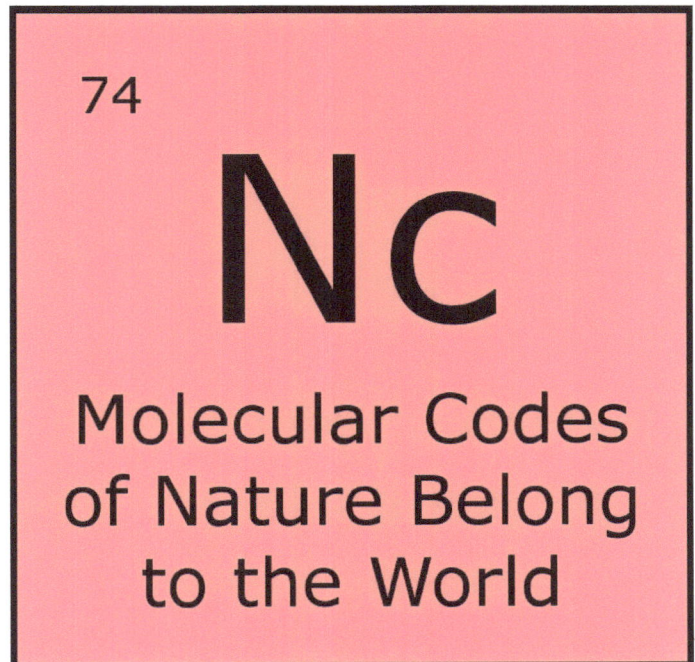

74
Nc
Molecular Codes of Nature Belong to the World

21 **Wu** Waste Material Utilization and Valorization	22 **Sa** Molecular Self-Assembly	23 **Ru** Reduce use of Hazardous Materials	24 **Dg** Design Guidelines	25 **Aq** Aqueous and Biobased Solvents	26 **Ee** Energy and Material Efficient Synthesis and Processing	27 **Ib** Integrated Biorefinery	28 **E** Enzymes	29 **Bm** Benign Metabolites	30 **Sn** Sensors
39 **Op** One-Pot Synthesis	40 **Ip** Integrated Processes	41 **Gc** In-Situ Generation & Consumption of Hazardous Materials	42 **Cm** Computational Models	43 **Il** Ionic Liquids / Non-Volatile Solvents	44 **R** Renewable / Carbon-Free Energy Inputs	45 **C** Carbon Dioxide and other C1 Feedstocks	46 **Ac** Earth Abundant Metal Catalysis	47 **Md** Molecular Degradation Triggers	66 **Ex** Exposome
57 **Pi** Process Intensification	58 **As** Additive Synthesis	59 **Ch** C-H Bond Functionalization	60 **Ba** Bioavailability / ADME	61 **Sc** Sub- and Super-Critical Fluids	62 **Es** Energy Storage / Transmission Materials	63 **Sb** Synthetic Biology	64 **Ht** Heterogeneous Catalysis	65 **Dp** Degradable Polymers and Other Materials	48 **Co** In-Process Control and Optimization
75 **Ss** Self-Separation	76 **W** Non-Covalent Derivatives / Weak Force Transformation	77 **Is** Inherent Safety and Security	78 **Ts** High Throughput Screening (Empirical / In Vivo / In Vitro)	79 **S** "Smart" Solvents (Obedient, Tunable)	80 **V** Waste Energy Utilization and Valorization	81 **Bt** Biologically-Enabled Transformation	82 **Hm** Homogeneous Catalysis	83 **Pd** Prediction and Design Tools	84 **Ga** Green Analytical Chemistry

Green Chemistry and Green Engineering Elements

In order to address the humanitarian goals that are inextricably linked to preserving our natural world, invention and innovation are essential. While efficiency will be necessary, it will be far from sufficient. Efficiency will help you do the thing you are doing better; it will not help you do a better thing. Being more efficient in our use of unsustainable resources, products, processes, and systems is not going to fundamentally alter the unsustainable trajectory we are on. Doing something that is unsustainable more efficiently does not make it sustainable.

The design, invention, and innovation that is required is embedded in the Principles of Green Chemistry and Green Engineering. The scientific and technological breakthroughs using these principles are meant to change the nature and character of the material basis of our society and our economy – including the materials used to generate, store, and transport our energy – to make it healthful rather than toxic, renewable rather than depleting, and restorative rather than degrading. The manifestation of these principles are the discoveries that go from the laboratory to making a difference in the world.

This collection of science and technology is the manifestation of the Twelve Principles of Green Chemistry and the Twelve Principles of Green Engineering.

PREVENT WASTE

Regardless of its nature, waste consumes resources, time, effort, and money both when it is created and then when it is handled and managed at end of life, with hazardous waste requiring even greater investments for monitoring and control. Creating, handling, storing, and disposing of waste is necessarily an expense and does not add value in terms of innovation or performance. In processes of production, therefore, waste is always undesirable in any form.

Ideally, molecules, products, processes and systems would be designed to not create waste. We should aim to eliminate even the concept of waste by ensuring that all outputs are feedstocks elsewhere, mimicking natural systems. Whether the waste is material, energy, space, time, or the derivative of all of these, money, there are design strategies that can and are being implemented in Green Chemistry and Engineering to address the issue at the most fundamental level: prevention.

21 **Wu** Waste Material Utilization and Valorization
39 **Op** One-Pot Synthesis
57 **Pi** Process Intensification
75 **Ss** Self-Separation

Waste Material Utilization and Valorization

21
Wu
Waste Material Utilization and Valorization

Waste is a human-centered concept. In Nature there is, for all practical purposes, no waste. In Nature, organisms and geo-systems evolve to utilize the "waste" of one process to nourish, sustain, and strengthen another. While the history of human-designed chemical systems has been linear (e.g., take-make-waste), there is a recognition that genuine elegance requires the building of circular systems where materials and energy flow in cycles as they do in Nature. This depends on developing the tools, techniques, and approaches for "designing waste" such that the "waste" itself is considered an additional product of the system.

These approaches of chemical waste valorization have found large-scale application in a variety of settings from individual product manufacturing processes to entire mega-scale factories to urban and regional networks. Transforming waste into a value-added product is a science in its infancy and it will need to develop considerably in sophistication and scale in order to displace the wasteful linear processes that have historically dominated.

"One-Pot" Synthesis

When chemical transformations have taken place historically, they have often required many steps, especially for complex molecules such as those in the field of pharmaceuticals. Each time there is a multi-step synthesis, there is the need for separation, isolation, and purification that results in loss of material, increased energy usage, and time lost.

Where possible, chemists should aim to design "one-pot" synthesis, in other words, transformations that can be carried out in a single reaction vessel without isolation, purification, or other wasteful steps. These more efficient processes must also avoid potential tradeoffs such as relying on more toxic substances or generating lower quality products. When implemented carefully, fewer steps and fewer reaction vessels can provide significant benefits in efficiency and waste reduction.

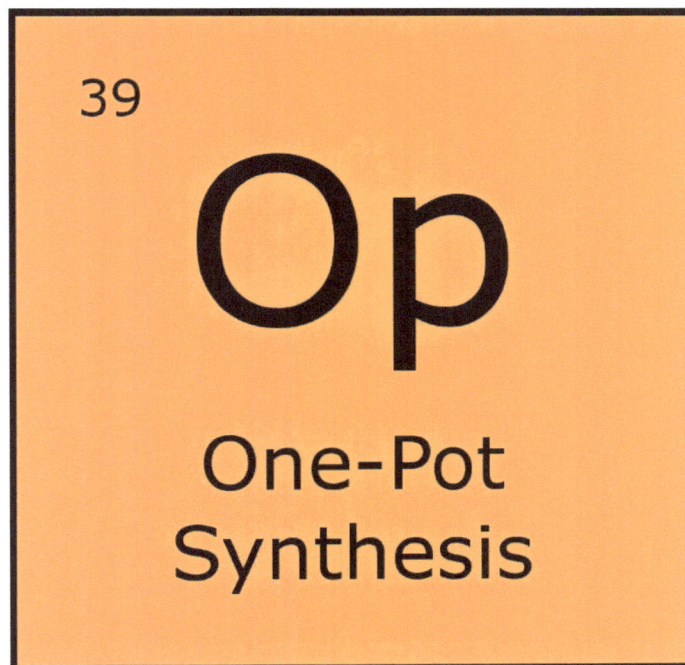

39

Op

One-Pot
Synthesis

57 Pi

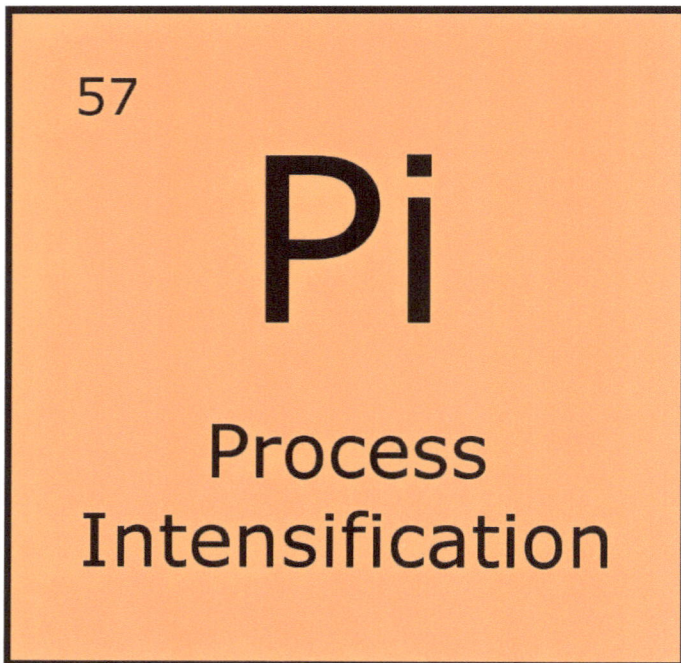

Process Intensification

Process Intensification

Traditional methods of chemical manufacturing have intrinsic inefficiencies that result in wastes of materials, energy, time, and space. Through thoughtful redesign, entire manufacturing processes that had been split into large numbers of "unit processes" of reaction, separation, purification, etc., can be combined. The benefit is not merely waste reduction but also lower capital and operating costs while increasing inherent safety. The concept of having small continuous-flow "reactors" that can be increased in number, so-called "numbering up," can realize these advantages compared to conventional "scaling up" of bulk processes.

Self-Separation

The chemicals and materials enterprise is one of the major consumers of energy in industry. One very large piece of this energy consumption comes from separation processes, encompassing isolation, purification, and some kinds of cleaning. In addition to the energy that is directly input into traditional separation systems, the dependency of these processes on large volumes of solvents also consumes significant amounts of embedded energy from solvent manufacturing and purification.

Systems designed to facilitate "self-separation" can decrease energy and material usage when developed properly. Designing a molecular product to separate from its reaction matrix or enabling an impurity generated in a process to self-separate due to intrinsic factors can have sustainability advantages.

75

Ss

Self-Separation

Atom Economy

The modern miracle of chemical enterprise is the ability to transform the materials that occur naturally in the world into new materials with new properties and performance that would not otherwise exist. History has shown there are thoughtful and wise ways to engage in molecular transformation and there are methods that are profoundly toxic, wasteful, and depleting. The new emerging synthetic methods that have been demonstrated in the field of green chemistry come from thoughtful life-cycle design and often use Nature and biological systems as an inspiration, mentor, and guide. Through this type of thoughtful design, we move from a narrow definition of mere efficiency to one of holistic productivity.

22
Sa
Molecular Self-Assembly

40
Ip
Integrated Processes

58
As
Additive Synthesis

76
W
Non-Covalent Derivatives / Weak Force Transformation

Molecular Self-Assembly

22

Sa

Molecular
Self-Assembly

The process of intentional bond-making has been one of the most important accomplishments of chemistry. Historically, this process has often been forced through the use of energy or reactive reagents to occur at the time, place, and rate that is desired. The more complex the molecule, the more steps and energy that has often been employed.

Looking at natural and biological systems, there are many examples where complex molecules with extensive stereochemistry and intricate ring systems are realized through molecular self-assembly that takes place upon appropriate stimulus. This can be achieved in human-designed synthesis as well. It is, of course, always important to not simply design a self-assembling molecule by shifting the energy and reagent inputs to another part of the product life-cycle such as its precursors.

Integrated Processes

Long chains of unit reaction processes accumulate energy and material waste due to the separation, isolation, and purification steps usually encountered after each step. These transitions are among the most energy-intensive aspects of an overall process. There are almost always material losses due to transfer from a reactor to another process step. By designing processes to be integrated to the highest feasible degree, there can be significant advantages not only for energy and material efficiencies, but also for the time saved and reduced exposures to chemical workers.

It can be valuable to integrate not only individual processes but numerous processes as well to accomplish goals such as utilization of waste heat/material from one process as a feedstock for another.

40

Ip

Integrated Processes

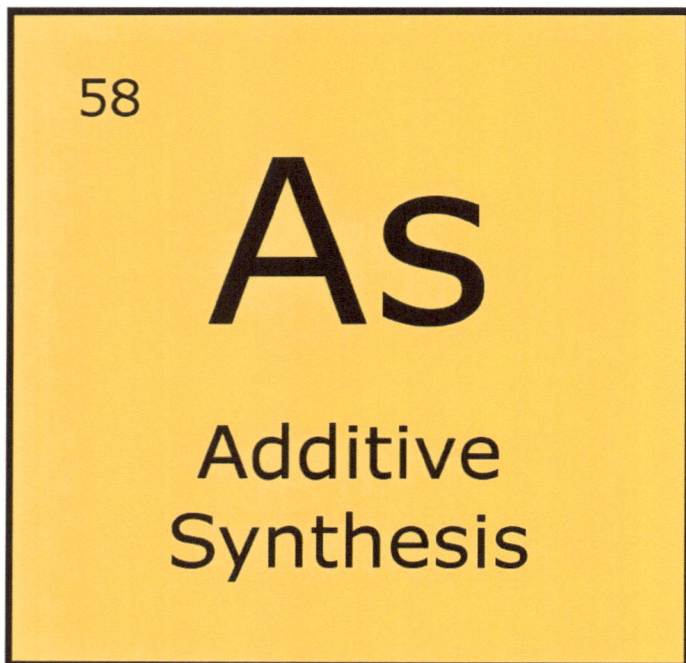

Additive Synthesis

58
As
Additive Synthesis

Additive Synthesis

The use of addition reactions can significantly decrease intrinsic waste generated by the synthetic method. By building up a molecule atom by atom or fragment by fragment, chemists use only what is necessary in constructing the target. It is essential to be thoughtful about not simply shifting the burden of waste from one step to another step or another part of the process, but wisely employing addition reactions has benefits over elimination and substitution reactions that inherently generate waste as part of the nature of the reaction.

Conceptually, the addition reaction can be translated beyond the molecular scale. For example, additive manufacturing is similarly beneficial for larger scale material assembly and applications.

Non-Covalent Derivatives / Weak Force Transformation

Chemists have pursued the mastery of bond-making with for more than 200 years with great success. However, we should note that Nature accomplishes impressive functional modifications and performance not only through covalent bonds but also extensively using weak-force interactions. These weak forces engage at the time and place necessary to impart useful properties precisely when and where they are needed. Weak forces guide many bond-forming pathways in Nature.

While humans have been increasingly aware of the essential importance of weak forces, there is a need for greater mastery of these powerful tools such that they can be used as design levers to achieve function and performance with minimal harm to the environment.

76

W

Non-Covalent Derivatives / Weak Force Transformation

Less Hazardous Synthesis

While efficiency has historically served as a proxy for sustainable practices in the chemical enterprise, it is imperative that the goal of reducing the quantity of material and energy consumed is closely coupled with considerations related to the nature of that material and energy. Using less may not have the beneficial effects of reducing the overall hazard of the synthetic process depending on the nature of the feedstocks, reagents, and auxiliary chemicals. It is imperative that these inputs and outputs, in addition to the intended product, are as inherently benign as possible. Further, the conditions under which synthetic processes are carried out should be considered when pursuing the goal of a more sustainable chemical enterprise. This offers benefits from environmental and human health perspectives in addition to a reduction in vulnerability to chemical accidents and sabotage.

23
Ru
Reduce use of Hazardous Materials

41
Gc
In-Situ Generation & Consumption of Hazardous Materials

59
Ch
C-H Bond Functionalization

77
Is
Inherent Safety and Security

23
Ru
Reduce use of Hazardous Materials

Reduce Use of Hazardous Materials

It is more impactful to design for reduced hazard than for controlling exposures. The use of hazards – physical (e.g., corrosivity, reactivity, explosivity, flammability), toxicological (carcinogenicity, reproductive and developmental including endocrine disruption, neurological), or global (e.g., ozone depletion, greenhouse gases) – can be minimized or eliminated throughout the entire life-cycle of a chemical process. The approach applies to feedstocks, reagents, solvents, catalysts, byproducts, and the products themselves. New or alternative chemicals should be rationally designed for reduced hazard or, at a minimum, assessed and understood to have reduced hazard by comparison to materials being substituted, to minimize these negative environmental and human health consequences.

In-Situ Generation and Consumption of Hazardous Materials

Reactivity is an essential part of chemical transformations and is also closely associated with the hazardous nature of many substances. One strategy that can be employed to avoid exposure of workers or nearby communities to toxic/reactive reagents, is to generate and consume them in the processes without any significant accumulation of these substances. Through "in-situ" generation, the substance needs only to be generated in minuscule quantities for short amounts of time as it is needed before it has consumed that reaction.

41

Gc

In-Situ Generation & Consumption of Hazardous Materials

C-H Bond Functionalization

Organic chemistry creates new molecules by forming carbon-carbon (C-C) bonds. Most processes use raw materials with carbon-hydrogen (C-H) bonds and, to avoid side products, making the desirable connections often takes an indirect, roundabout route to avoid side products. The conventional chemistry toolbox includes substitution and elimination reactions that generate waste at the molecular level, adding to the process complexity as well as environmental burdens. Direct C-H activation through catalysis enables addition or rearrangement mechanisms that are inherently efficient. Continuous improvement in this area of research gives chemists more options for cleaner chemistry without sacrificing control over the reaction or limiting access to interesting chemical structures.

Inherent Safety and Security

Chemists and chemical engineers know the properties, structures, and conditions that underlie explosivity, flammability, and corrosivity. Designing molecules such that they are not capable of ignition, conflagration, combustion, or explosion is something that is within the skillset of chemistry today.

While safety has been a concern of the chemical enterprise for much of its history, this has often been accomplished through protection from the consequences of these physical hazards when they occur rather than through molecular design for reduction of intrinsic hazard. Upfront design gives better protection from accidents as well as terrorism and sabotage vulnerability.

77

Is

Inherent Safety and Security

MOLECULAR DESIGN

The basis of our society and economy is synthetic chemicals and materials. While there have been significant advances in toxicology associated with identifying, and in some instances predicting, commercial chemicals that are likely to cause harm to human and ecosystem health, the gains in informing the a priori design of chemicals with reduced hazard to humans and the environment have been elusive. To realize the goal of designing chemicals that are safe and functionally relevant, there is a need to create an interdisciplinary body of knowledge at the nexus of computational chemistry, mechanistic toxicology, and big data analytics among others. It is only when we change the inherent nature of the chemicals and materials that are foundational to quality of life that we can truly advance toward a sustainable future that is no longer reliant on costly regulatory and technological controls of circumstances in which hazardous chemicals can be used and managed.

24
Dg
Design Guidelines

42
Cm
Computational Models

60
Ba
Bioavailability / ADME

78
Ts
High Throughput Screening (Empirical / In Vivo / In Vitro)

24

Dg

Design Guidelines

Design Guidelines

Design is intentional. If a chemical contains a hazard that is not intended, it is a design flaw. Yet, many of our made-made chemicals contain hazards to humans or the biosphere by accident or lack of thoughtful design.

As we understand the underlying basis of hazard to human health and the environment at the molecular level, we can design to avoid it. As we understand the properties which enable chemical accidents and other adverse consequences, we can and must design to avoid these outcomes where possible.

Computational Models

With increasingly deeper molecular-level understanding of the nature of chemical hazard comes increased potential for the use of computational models to assess, predict, and design out hazards from the chemicals we use and produce. The wide range of physical/chemical properties that are the underlying basis for toxic mechanisms of action, exposure pathways, and transport and fate can be evaluated in-silico with greater accuracy and the insights derived from these evaluations can be used to avoid the adverse consequences that have marked the less-desirable aspects of the history of the chemical enterprise.

42

Cm

Computational Models

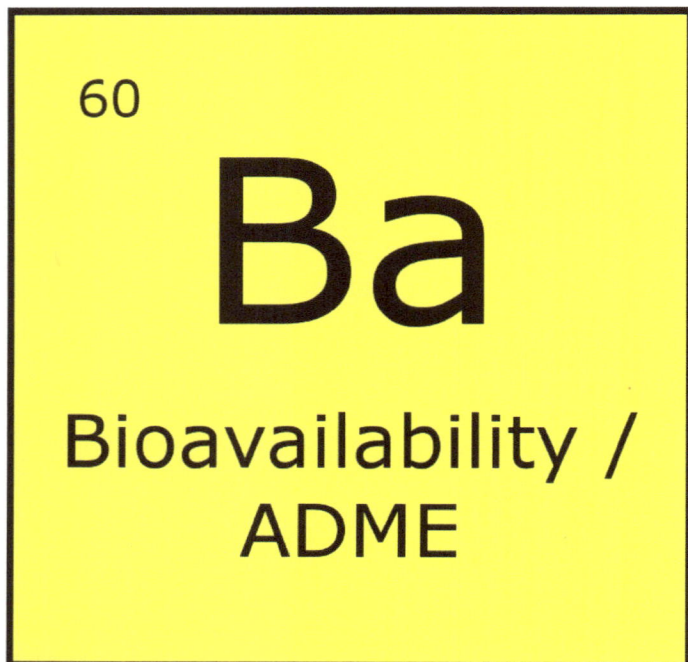

Bioavailability / ADME

60
Ba
Bioavailability / ADME

Bioavailability / ADME

At the foundation of protecting living things from toxicological impacts is the concept of ADME (absorption, distribution, metabolism, excretion), describing how chemicals behave in and are changed within an organism. By understanding the molecular parameters that control the stages of ADME, chemists can control a chemical's bioavailability—its ability to access a biological system. The insights from pharmaceutical research in trying to maximize bioavailability have provided intellectual tools that can be used in all aspects of the chemical enterprise in thoughtful design for hazard reduction.

High Throughput Screening
(Empirical / In Vivo / In Vitro)

Empirical testing of the toxicity of a chemical is an essential part of the deep understanding of the potential consequences of a chemical. As we move away from animal testing due to ethical and financial drivers, the development of bioassays is increasingly important. Emerging bioassays cover a large number of biological processes and toxicological endpoints and can be done within a high throughput framework. The extensive amounts of data generated provide insights into the concerns that may be associated with specific chemicals, and in aggregate have potential to inform design guidelines for safety of future chemicals.

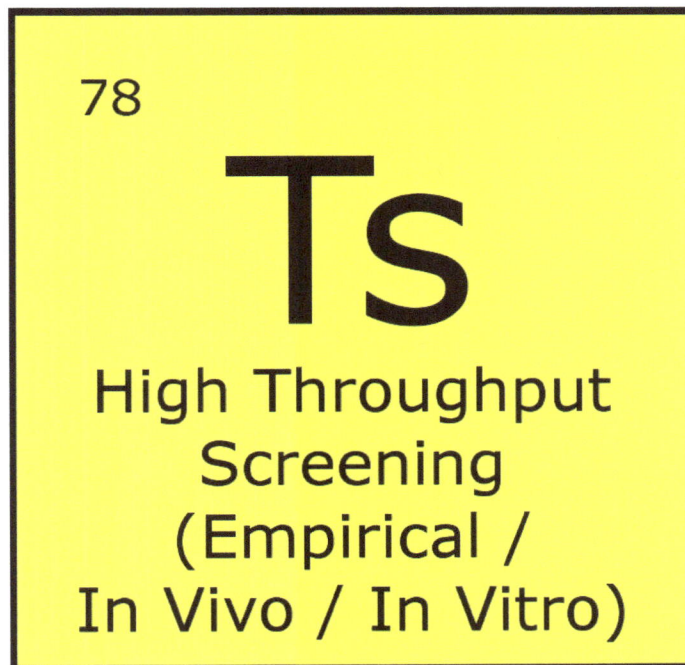

78

Ts

High Throughput Screening (Empirical / In Vivo / In Vitro)

Solvents/Auxiliary Chemicals

The cost, environmental impact, and safety of a chemical processes are often driven by the solvents and other auxiliary chemicals. The amount of solvent used often exceeds raw materials, reagents, and products, particularly in the case of separation and purification processes. Once again, while quantity of these chemicals is an important consideration from a perspective of material and energy efficiency, it is the nature of our historic solvents that has posed the greatest challenge to the environment and human health. Conventional solvents have generally been volatile, increasing likely exposures; hydrophobic, serving as long-term sources of concern in environmental systems; and toxic to ecosystems and humans, particularly notable in terms of worker exposure. Just as we aspire to design safer chemical products, the same effort and attention should be brought to bear on the design of safer solvents or solvent-free processes through both improving the inherent nature of these compounds as well as minimizing their exposure potentials.

25
Aq
Aqueous and Biobased Solvents

43
Il
Ionic Liquids / Non-Volatile Solvents

61
Sc
Sub- and Super-Critical Fluids

79
S
"Smart" Solvents (Obedient, Tunable)

Aqueous and Biobased Solvents

25

Aq

Aqueous and Biobased Solvents

Water is the original solvent: the solvent of life. And yet, the long history of the chemical enterprise largely eschewed water as a solvent. As petrochemical processes came to be dominant, it was considered logical that organic solvents would be most compatible with hydrocarbon-based transformations. Research launched in the 20th century but accelerating in the 21st century has shown that water as a solvent is desirable for many chemical processes, even those where it was once thought not possible. Water has been shown in cases to accelerate reaction rates and enhance selectivity. The performance gains are in addition to benefits such as non-toxicity, non-flammability, cost-effectiveness, and ready availability in most instances.

Ionic Liquids / Non-Volatile Solvents

When salts can be designed to perturb their crystal structure adequately such that they are a liquid at room temperature, they are known as room temperature ionic liquids (RT-IL. These liquids have been demonstrated to be effective solvents in a wide range of applications while having negligible vapor pressure. This is important as lack of vapor pressure avoids one of the most concerning routes of exposure for most volatile solvents, that of respiration. Another major concern with traditional solvents is the impact on atmospheric chemistry and air pollution. With RT-IL, this type of atmospheric transport is not possible.

While the lack of vapor pressure is an elegant attribute for this class of solvents, it is also critical that these solvents are designed for low toxicity.

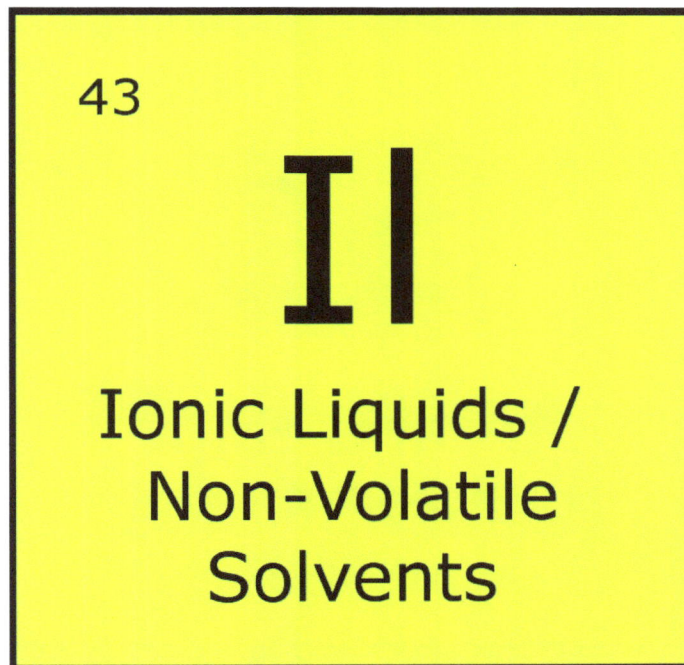

43

Il

Ionic Liquids / Non-Volatile Solvents

61

Sc

Sub- and Super-
Critical Fluids

Sub- and Super-Critical Fluids

There are gases that, when subjected to a certain temperature and pressure condition, known as their critical point, become materials that are neither gas nor liquid but rather "super-critical" fluids. While these fluids have been known for centuries, recent discoveries have renewed interest in their excellent solvent properties. Carbon dioxide, both in a sub- and super-critical state, has been demonstrated at large scale as a practical solvent for everything from synthesis to extraction, cleaning, and analysis.

The green advantages of supercritical fluids are numerous including low toxicity (for water and carbon dioxide), lack of flammability, tunability, and the possibility of "infinite recyclability" by cycling pressure. The use of carbon dioxide as a solvent does not require the generation of new CO_2 and does not necessarily contribute to greenhouse gas emissions.

"Smart" Solvents (Obedient, Tunable)

The ability to make a solvent respond to stimuli and change its properties under new conditions can be extremely consequential. Historically, solvents were energy consuming because solubility was controlled almost exclusively through heating and cooling. With next generation solvents that have been demonstrated and developed, solvents can be controlled by factors such as pressure or pH.

These new so-called obedient solvents open up possibilities for niche and industrial uses that change the energy profiles for many chemical processes.

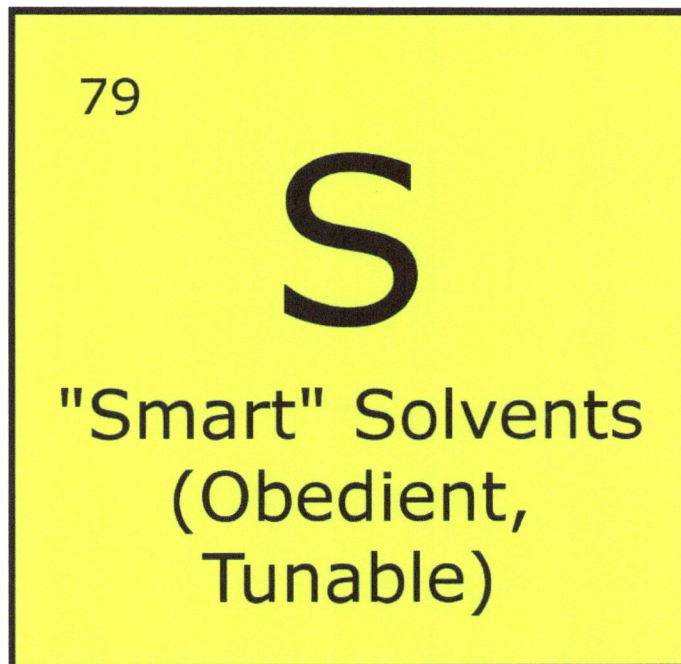

79

S

"Smart" Solvents (Obedient, Tunable)

ENERGY

The chemical sector consumes approximately 20% of total industrial energy consumption in the U.S., and contributes in similar proportions to U.S. greenhouse gas emissions. Given the reliance on fossil fuel resources and the associated greenhouse gas emissions, there is a clear indication that, at a minimum, the chemical enterprise should strive to be as energy efficient as possible, normalizing to chemical function rather than mass in the assessment. Gain in efficiency can be realized by considering both the quantity and quality of energy inputs as well as waste energy utilization. Of course, the chemical sector has a significant role to play in changing the nature of our energy feedstocks toward ones that are renewable, and developing materials that can enhance energy generation, storage, and transmission to enable the use of these renewable energy sources.

26
Ee
Energy and Material Efficient Synthesis and Processing

44
R
Renewable / Carbon-Free Energy Inputs

62
Es
Energy Storage / Transmission Materials

80
V
Waste Energy Utilization and Valorization

Energy and Material Efficient Synthesis and Processing

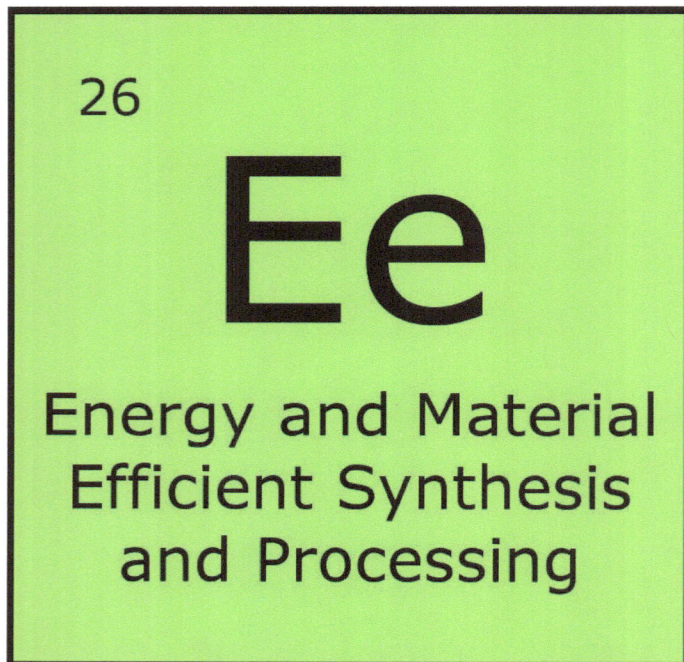

26

Ee

Energy and Material Efficient Synthesis and Processing

Synthesizing, transforming, and manufacturing raw materials into the desired chemical products requires the input of energy to drive the reaction. The form of this energy, in addition to the amount, has a significant impact on the environmental cost of carrying out the reaction. There have been recent advances to reduce the overall energy demand and subsequent environmental and economic impacts of chemical production by exploring different means of energy delivery including mechano-, electro-, photo-, and electromagnetic chemical-driven transformations. These alternative approaches can affect the amount of energy used in removal of contaminants (purification), retrieving the desired molecules (isolation), removal of extraneous/undesirable materials (cleaning), or breaking a complex mixture into its components, including the removal of water (separations).

Renewable / Carbon-Free Energy Inputs

As long as the chemical enterprise is among the largest energy consumers of the economy, there is a responsibility to ensure that the energy consumed is renewable and carbon-free. The options are becoming plentiful with solar and wind becoming a larger share of the energy grid. However, it must be recognized that all current sources of energy are not created equal in terms of ability to be applied to manufacturing processes of all chemicals. There will be needed advances in realizing renewable energy systems that can deliver the kind of scale and quality demanded for chemical processing, refining, and distillation.

44

R

Renewable / Carbon-Free Energy Inputs

62

Es

Energy Storage / Transmission Materials

Energy Storage and Transmission Materials

Even the most sustainable and renewable sources of energy risk going to waste when not utilized immediately if there is not adequate ability to store or transport this energy. Physical, mechanical, chemical, and other methods of energy storage and transmission will require the underlying materials to be benign and Earth abundant. Technologies such as batteries or hydrogen storage will only be as sustainable as the materials used throughout their life-cycles.

Waste Energy Utilization and Valorization

Waste materials are highly noticeable when they fill up landfills around the world or become pollution in the ocean, but much less attention is paid to waste energy generation. Waste energy in forms such as heat, light, vibration, or noise has a cost due to the impacts of energy generation for which there is no return as well as the impacts on the immediate surroundings (e.g., warming, wear and tear, mechanical erosion). The ability to capture this waste energy and find value-added uses is an important part of any design strategies for a sustainable product process or system.

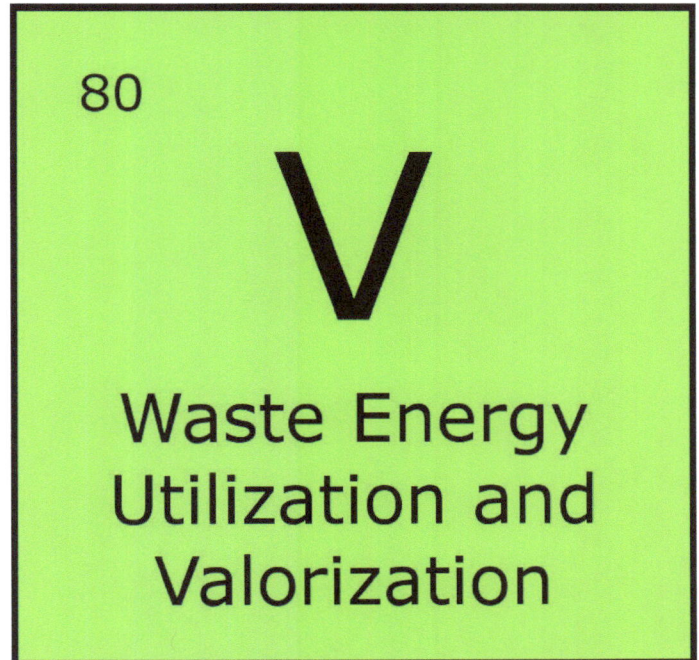

80

V

Waste Energy Utilization and Valorization

Renewable Feedstocks

Fossil fuel is the basis of the chemical, material, and energy foundation of the global economy. Fossil fuel feedstocks are used to generate electricity, to produce transportation fuels, and to produce a wide range of consumer goods, such as plastics, healthcare and drug products, and agrichemicals. These reserves are finite and pose additional challenges related to geopolitics and physical accessibility.

Given this context, there have been emerging and important efforts to increase the use of bio-based feedstocks for energy, chemicals, and materials production. Using renewable feedstocks from agricultural, forestry and aquatic resources, particularly residue and waste streams from processing these materials, will be essential to changing the material and energy basis of our economy and society. However, this must be implemented in a context of competition with food, land and water use, as well as benign and efficient downstream processing for recovery of the full value of the feedstock.

27 **Ib** Integrated Biorefinery

45 **C** Carbon Dioxide and other C1 Feedstocks

63 **Sb** Synthetic Biology

81 **Bt** Biologically-Enabled Transformation

Integrated Biorefinery

27	
Ib	
Integrated Biorefinery	

While matter is neither created nor destroyed, the molecular conversion of natural resources through industrialized processes has the ability to transform these resources from benefits into burdens. The conversion of energy-dense hydrocarbons to heat-trapping greenhouse gases is a high-profile example but the concept extends to resources like water, soil and minerals, and biodiversity. In the absence of the ability to fully renew and restore, there can be no license for our chemical enterprise to access and transform the parts of the geosphere and biosphere that cannot be replaced.

Perhaps the most materially efficient technological process in history is the refinery. The modern petrochemical refinery has lessons for the bio-based chemical economy that has not reached the same level of efficiency. Being able to retrieve value from all extractable fractions from high-volume, low-value to low-volume, high-value in an integrated biorefinery will be an important piece of realizing the goal of a bio-based economy.

Carbon Dioxide and other C1 Feedstocks

Carbon has been lusted over in its diamond form. Carbon has driven economies in its coal form. Carbon has emerged as a 21st century energy source in its methane form. Carbon is deadly in its carbon monoxide form. Carbon is essential to the biosphere and threatens climate stability in its carbon dioxide form. Carbon is poised to play a critical role in the evolution of sustainable chemistry. The emergence of C1 chemistry is the essential manipulation of one-carbon building blocks to transform them from recalcitrant to useful, from destructive to value-adding, from a challenge to opportunity.

Learning to sustainably access the promise of carbon dioxide and other C1 molecules as feedstocks is a critical frontier in green chemistry.

45

C

Carbon Dioxide and other C1 Feedstocks

Synthetic Biology

New genomes are no longer science fiction. The ability to design and control biological organisms and processes is well on its way to becoming one of the most powerful tools to benefit the move toward a sustainable world. It also has the potential to cause almost unimaginable harm. Thoughtful development, use, and implementation of synthetic biology processes must proceed within a sustainability construct. Long-term implication of use and misuse must be considered and built in prior to implementation. The impact of synthetic biology will be felt by this world in the coming decades and it is essential that these impacts be positive and not follow the errors of previous technological advances of realizing a specific utility coupled with myriad unintended consequences.

63

Sb

Synthetic Biology

Biologically-Enabled Transformation

Nature is the original chemist carrying out an inconceivable number of chemical transformations each second with a volume that dwarfs the combined output of every chemical company and an elegance that puts the occupants of the best chemistry departments to shame. Capturing and harnessing the power of biological processes to carry out chemical transformations can be a powerful strategy for a sustainable economy. Oriented toward renewable, bio-based feedstocks, these processes (e.g., fermentation) can be utilized for production of materials in a manner that is supportive of and conducive to life.

81

Bt

Biologically-Enabled Transformation

CATALYSIS

There are few areas of chemistry that exemplify sustainability better than catalysis. Catalysis allows for increasing the rate of a transformation, increased efficiency, use of less feedstock, enhanced product quality, lower waste, and lower emissions while at the same time increasing the profitability of a process or product. Virtually every major petrochemical, specialty chemical, or pharmaceutical company would not be economically viable without the use of catalysis. The blend of environmental benefit and economic returns via catalysis makes it among the most obvious and powerful tools to advance sustainability through chemistry.

28
E
Enzymes

46
Ac
Earth Abundant Metal Catalysis

64
Ht
Heterogeneous Catalysis

82
Hm
Homogeneous Catalysis

28

E
Enzymes

Enzymes

Enzymes are often viewed as the pinnacle of what catalysis strives to be. They carry out conversion of chemicals with an elegance and selectivity that most chemists would only dream of. Yet there is room to improve on the current limitations such as tolerance to a wide-range of conditions including temperature, pressure, pH, and concentration gradients. The development of manufacturing methods that exploit the advantages of enzymes will be an important contribution to sustainable chemistry manufacturing.

Earth Abundant Metal Catalysis

Historically, some of the most toxic and/or precious metals were often developed and used as catalysts and targeted for research and study. These include osmium, mercury, chromium, platinum, palladium, gold, and iridium. The cost and sustainability impacts of this approach are significant. While it must be noted that there exist excellent recovery schemes, each of these have costs and risks that should be minimized.

The exploitation of Earth-abundant metals (e.g., iron, copper) that are not scarce nor depleting is an active area of development that will be important to achieve the necessary advantages of metal catalysis without the same associated historic impacts.

46

Ac

Earth Abundant
Metal Catalysis

Ht

64

Heterogeneous Catalysis

Heterogeneous Catalysis

Heterogeneous implies that something is not uniform in nature and contains a few or many different components or substances that make up the whole. In chemistry, this includes mixtures of materials in different phases (gas, liquid, solid). Considering catalysts important in sustainable chemical pathways, heterogeneous systems confer many advantages compared to homogeneous catalysts including better stability, ease of handling, separation, and simplified recycling of the catalyst.

Homogeneous Catalysis

When something is the same throughout and consists of all the same parts it is homogeneous. In chemistry, this can refer to uniformity of composition, but also mixtures of materials in the same state of matter (solid, liquid, or gas). Homogeneous catalysts, in which the catalyst and reactant are in the same phase, are at the heart of extremely important methods in green chemistry. Due to the nature of homogeneous catalysts and the ability to easily characterize how they function, they are relatively easy to design at a mechanistic level for enhanced performance and can positively impact health, leading to improved environmental outcomes.

82

Hm
Homogeneous
Catalysis

DEGRADATION

Persistent substances may remain in their natural and man-made environments for an indefinite time. These compounds may accumulate to reach levels that are harmful to health, environment, and natural resources. Such contamination may be poorly reversible or even irreversible, and could render natural resources such as soil and water unusable far into the future. As such, there is a need to design chemicals and materials, particularly those that are intentionally or unintentionally distributed in the environment, that are stable during their useful life, and then degrade once they are no longer functionally necessary. Designing for degradation is necessary but not sufficient to realize the goals of sustainable chemistry; the subsequent degradation products should, themselves, be benign to human health and the environment.

29 **Bm** Benign Metabolites
47 **Md** Molecular Degradation Triggers
65 **Dp** Degradable Polymers and Other Materials
83 **Pd** Prediction and Design Tools

29

Bm

Benign Metabolites

Benign Metabolites

Metabolites are formed by degradation or transformation of chemicals in a living organism. Metabolic processes in humans, plants, or any living organism are relevant to sustainable chemistry. Metabolites may have beneficial functions within an organism to regulate various processes, or they may be inert. However, some can cause toxicity or adverse health effects. Awareness of these process is important as sustainable design must anticipate product breakdown and not just the product itself in order to minimize harm to humans and the environment. Benign parent compounds can produce harmful metabolites. This is especially relevant to biologically active chemicals such as pharmaceuticals and herbicides or pesticides that where small changes to chemical structure during physiological processes could lead to unintended effects.

Molecular Degradation Triggers

Molecular degradation triggers are events that strategically promote breakdown of molecules. These could include chemical- or pH-sensitive functional groups or UV light triggers. Today there are many stable compounds that persist in the environment far beyond the desired lifetime of the product. Molecules incorporating triggers could help prevent harmful impacts that arise from chemical persistence, if designed to form benign by-products following the intended end of life of chemicals and products.

47

Md

Molecular Degradation Triggers

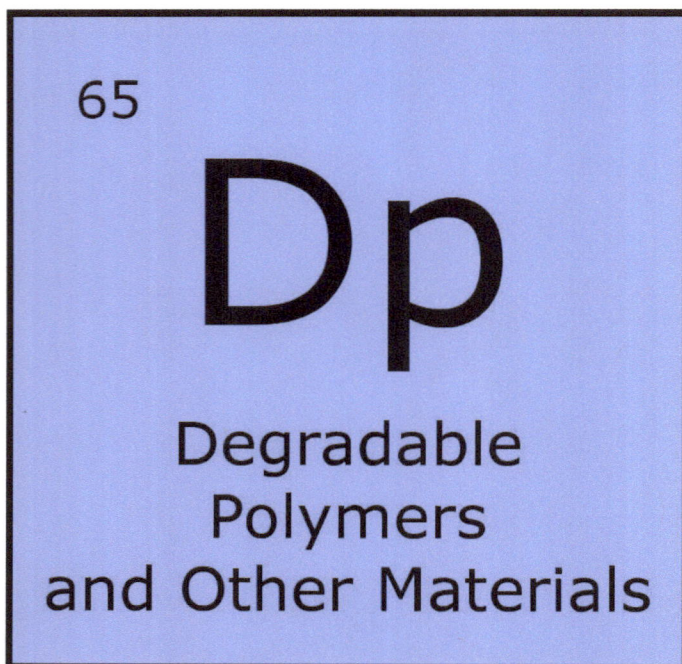

Degradable Polymers and Other Materials

65

Dp

Degradable Polymers and Other Materials

For a chemical material to endure and persist beyond intended use serves little purpose and has been the cause of serious problems. Polymers should not be (effectively) immortal. Yet, many materials approach that level of persistence when compared to the human time-scale. This has resulted in well-known concerns for plastics in landfills and in the oceans. It also is of concern for lesser-known issues such as the potential impacts of persistent microplastics or water-soluble synthetic polymers that are being integrated into ecosystems and the biosphere. Polymers can be designed such that their entire lifetime is approximately the same as their useful lifetime.

Prediction and Design Tools

Designing next generation chemicals and materials will require next generation prediction and design tools. Analytical understanding and computational models continue to be more powerful, allowing increased insights to go beyond simply predicting potential adverse impacts toward informing design. This is critical because predicting that something may be harmful without a mechanistic basis for that model outcome does not provide sufficient and necessary guidance to a chemist on how to redesign the molecule to preserve function while minimizing or eliminating the hazards.

83

Pd

Prediction and Design Tools

MEASUREMENT AND AWARENESS

To know whether an action or invention is positive or negative for the world, there needs to be some kind of feedback from the system. In the absence of this feedback, there is little more than insufficient guesses and suppositions. However, the way that we choose to measure the performance and impact of our products, processes, and systems will often have a profound impact on the understanding that we take away from these measurements. (Even the compass is only two-dimensional.) If we are operating in multi-dimensional systems like those involved with sustainability, we will need measurement systems that are as multi-dimensional if we wish to gain insight from them. Traditional, limited measurements such as efficiency measures will be insufficient for this complex system. Additionally, our measurement systems have often been retrospective tools that would give us reports on what happened in the past, sometimes with long enough lag times that the information itself is irrelevant. We now have tools and computational abilities that can provide real-time insights and awareness that allow for relevant action.

30
Sn
Sensors

66
Ex
Exposome

48
Co
In-Process Control and Optimization

84
Ga
Green Analytical Chemistry

| 30 |
| Sn |
| Sensors |

Sensors

Chemical and chemo-electronic sensors can now provide a level of detection and real-time awareness of factors and dynamics that were once thought unknowable. With applications ranging from air and water quality to health indicators to environmental ecosystems shift, the development of ubiquitous, integrated sensors can provide data that inform insights into planetary and human health. These same sensors can inform and empower more efficient use of materials and energy in manufacturing and consumer use. These approaches will be critical to maintaining a sustainable society.

In-Process Control and Optimization

In manufacturing processes and operations, the way to verify smooth operation was conventionally through periodic analysis, requiring extractive testing and analysis of the product, often destructively. This approach was time- and labor-intensive and led to waste materials and energy. Often, product that did not meet specifications (so-called off-spec) would need to be discarded as waste, frequently in large quantities. It is now possible and important that future processes and even operations beyond manufacturing utilize in-process control and optimization through the use of real-time sensors and analysis. In this way, there is the ability to sense off-spec material before it is formed in significant quantities and to make the appropriate corrections in-situ. This becomes especially vital when pursuing the goal of process integration.

48
Co
In-Process Control and Optimization

Ex

66

Exposome

Exposome

The exposome refers to everything an individual is exposed to beginning in utero through end of life. Each of these exposures can accumulate, compound, and interact with each other and affect one's health. Exposures include anything from one's environment, diet, workplace, and lifestyle. An individual's unique physiological characteristics and genetics can play a major role in determining the effects of exposure. In terms of sustainability, it becomes important to consider the potential exposures in the design stage for any product or chemical. By better understanding the exposome we can make design choices that improve safety and minimize harmful exposures.

Green Analytical Chemistry

Analysis of the air, water, and land to detect pollution was historically conducted with field sampling protocols required and extensive effort to transfer materials to specialized laboratories where extensive work-up generated large volumes of solvents and waste. The processes of measurement and analysis of an environmental problem often contributed to other environmental problems. With the use of real-time, in-field analysis, the necessary measurements can be taken without the wasted time, material, and energy. This "green analytical chemistry" has been extended to a wide range of analytical techniques beyond environment analysis, reducing the quantity of materials needed and waste generated.

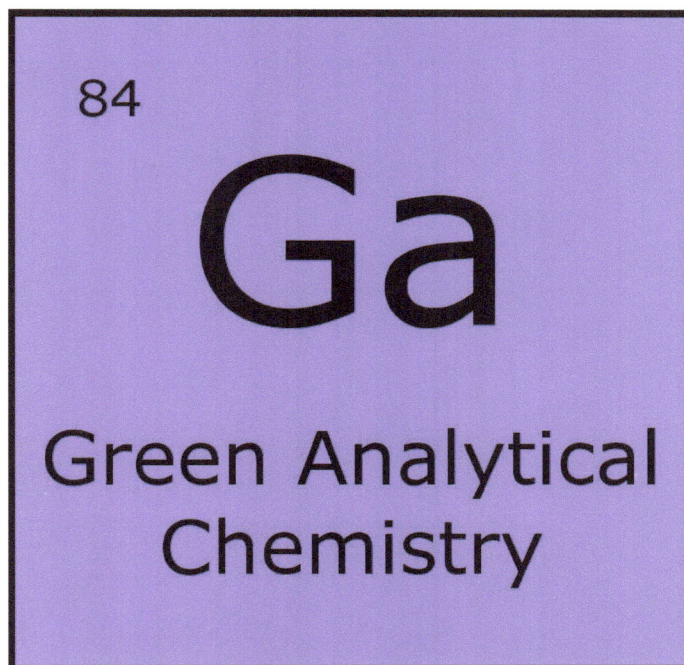

84

Ga

Green Analytical Chemistry

5 **B** Biomimicry	6 **Cb** Life Cycle Cost-Benefit Analysis	7 **Ae** Atom Economy	8 **Pr** Extended Producer Responsibility	9 **Ea** Epidemiological Analysis and Ecosystem Health
13 **Ce** Circular Economy	14 **Fc** Full Cost Accounting	15 **Ef** E-Factor	16 **Pb** Property Based Regulation	17 **Aa** Alternatives Assessment
31 **Bd** Benign by Design	32 **Hc** Harm Charge / Carbon Tax	33 **Ff** F-Factor	34 **Ct** Chemical Transparency	35 **Lc** Life Cycle Assessment
49 **Ie** Industrial Ecology	50 **Dc** Depletion Charge	51 **Ql** Qualitative Metrics	52 **Cl** Chemical Leasing	53 **So** Solvent Selection Screens
67 **Tg** Trans-Generational Design	68 **Rf** Sustained Research Funding	69 **Qn** Quantitative Metrics	70 **Se** Self-Enforcing Regulations	71 **Cf** Chemical Footprinting
85 **Be** Bio-Based Economy	86 **Ci** Capital Investment	87 **Bb** Chemical Body Burden	88 **I** Innovation Ecosystem Translation from Lab to Commerce	89 **Et** Education in Toxicology and Systems Thinking

ENABLING SYSTEMS CONDITIONS

Certainly, chemistry must not simply be a puzzle to be solved. Certainly, it's not simply an academic curiosity. It always has had the thread of purpose and impact to it. And because of the understanding that when we introduce new things into the universe that the universe has never seen, we need to anticipate the consequences. All of green chemistry and green engineering has been focused on those parts of the endeavor. But in order to make that endeavor real, in order to make those discoveries, those insights, those inventions, real, discoveries need to go beyond the realm of science or technology.

The science and the technology may be essential. Without the science, advancement may not be possible. But even with the discoveries, the inventions, and the innovations, advancement is not going to happen alone with only science and technology. We recognize its that chemical molecular transformations depend on different conditions: heat, pressure, light, stirring, mixing, whatever it is. We need to think even more broadly about the system conditions that are necessary in order to bring about the positive impact at societal scale, and what is the urgency that is going to empower a sustainable transition, a transition to a sustainable world.

Conceptual Frameworks

Conceptual frameworks provide a basis for considering the role of the chemical enterprise in advancing sustainability and the associated complexities.

Such frameworks provide a means to understand the relevant concepts as well as their relationships to one another.

Frameworks of this nature help to inform the overarching design of the role of chemicals and chemistry in contributing to the goals of a sustainable future.

5	**B** Biomimicry
13	**Ce** Circular Economy
31	**Bd** Benign by Design
49	**Ie** Industrial Ecology
67	**Tg** Trans-Generational Design
85	**Be** Bio-Based Economy

B
5

Biomimicry

Biomimcry

Nature has had a 3.5 billion-year head start on designing products, processes, and systems and it shows. Nature is indescribably brilliant in its ability to produce the widest range of chemicals and materials using locally-available starting materials, usually at ambient temperatures and pressures, completely eliminating the concept of waste, and doing it in a way that is conducive to life. In the face of this ingenuity, the only wise and sustainable strategy is to take Nature as a mentor and a guide in designing human-made products and systems. The lessons to be learned are limitless and the benefits to be had, essential.

Circular Economy

Circular Economy refers to closing the material loop. This means working to keep resources in use for as long as possible and requires both the producer and consumer to be aware of how they are utilizing resources. It can help to ensure waste and resources, at any point from production to end of life of a final product, are reused, repaired, or recycled in some capacity. This helps to decrease waste and becomes especially important in sustainability efforts, helping companies to reduce waste, save money, and work to ensure there is not depletion of finite resources. The concept of circular economy should be considered in all phases of the product life cycle and can not only help to save resources, but also decrease the amount of toxic waste output.

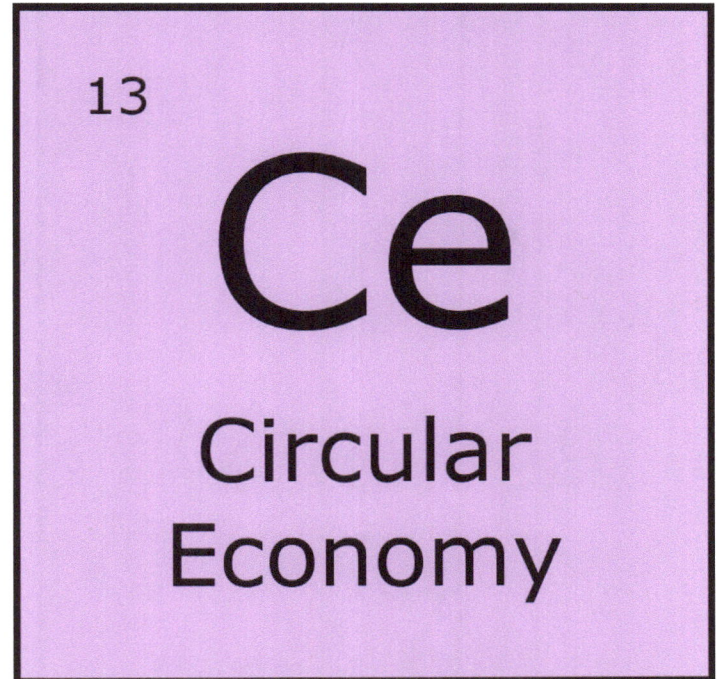

13

Ce

Circular Economy

Benign by Design

31

Bd

Benign by Design

Benign by Design

"Performance" of a product, process, or system has been a concept that has been narrowly defined in terms of how well something achieves a very specific goal. Perhaps it's a dye being a particular shade of blue or a lubricant's ability to reduce friction. While performance has been focused on the desirable qualities, this has often come with undesirable or adverse unintended consequences.

Benign-by-design builds all factors into the definition of performance such that inherent safety is also included along with function.

Industrial Ecology

Industrial ecology aspires to manage material and energy flows in the same way that an ecosystem performs, with the aims of ultimately decreasing environmental stress and improving resource efficiency. This can be done by considering industrial processes as a closed loop as much as possible, creating the opportunity to decrease inputs through the reuse of outputs either within a facility or between different organization. By viewing waste as a resource – by closing these material and energy loops – waste and resource depletion caused by industry can be minimized. Industrial ecology helps to find solutions to environmental problems by identifying problem areas in business and industry and can aid in improved sustainability practices.

49

Ie

Industrial Ecology

Trans-Generational Design

67
Tg
Trans-Generational Design

Trans-Generational Design

It has been shown that exposure to a chemical by one generation can have deleterious effects on future generations, even without direct exposure themselves. This is especially relevant to chemicals that interfere with the body's hormonal systems, leading to a variety of adverse health effects including reductions in sperm counts as well as tumors, birth defects, and other developmental disorders. In this way, the chemistry we practice and the chemicals we produce today can have lasting and debilitating effects on future generations. This emerging knowledge carries an enormous burden of responsibility to be accountable not only to today's population but to those not yet born.

Bio-Based Economy

Bio-Based Economy refers to the innovation in utilization of biomass to sustainably make bio-based products such as chemicals, materials, fuel, and energy. It is an attempt to move away from reliance on depleting resources and rely on renewable, bio-based feedstocks. Traditionally, chemicals have been made from petroleum feedstocks. Although chemical production only accounts for 5%–7% of petroleum consumption, petroleum sources represent over 98% of chemical feedstocks. The advantages for moving to renewable feedstocks include an opportunity for innovation and a chance to take advantage of nature's ability to perform exquisitely selective chemistry. Petrochemical feedstocks provide very simple hydrocarbons, which chemists have learned to make more complex. Natural feedstocks are inherently different. They are complex molecules, and chemists are still developing elegant ways to efficiently transform them into useful products.

85

Be

Bio-Based
Economy

ECONOMICS AND MARKET FORCES

Current global economic systems and market forces tend to drive perverse incentives that have resulted in the design and evolution of our existing chemical infrastructure.

These economic and market systems are critical and will either catalyze or retard the implementation of our science and our technology. It is important to recognize that in addition to traditional approaches of cost benefit analysis, full cost accounting, risk analysis, or traditional metrics, we also are going to need to understand factors including where investment models are made, what kind of business models and business frameworks will be either speeding up the adoption of Green Chemistry and Green Engineering or rejecting it no matter how transformative it may be, and how do we integrate these economic and market forces in order to accelerate and have our technologies adopted.

These powerful drivers can be harnessed to encourage behavior and decision-making that is aligned with the goals of a sustainable future.

6
Cb
Life Cycle
Cost-Benefit
Analysis

14
Fc
Full Cost
Accounting

32
Hc
Harm Charge /
Carbon Tax

50
Dc
Depletion
Charge

68
Rf
Sustained
Research
Funding

86
Ci
Capital
Investment

Life Cycle Cost-Benefit Analysis

Cost-Benefit Analysis is a decision-making process that helps to identify and quantify the costs and benefits associated with all potential scenarios or options. It is a systematic approach to estimating the strengths and weaknesses of alternatives used to make decisions between alternatives. Opportunity cost, the benefit missed by choosing one option over the other, is often also factored into the analysis. It can be used to compare intangible items, such as ecosystems services or the benefit of choosing an option that confers less environmental and negative health impact over a more detrimental option. In sustainability practices, cost-benefit analysis can help to identify options that may confer more favorable environmental conditions, while also minimizing costs over the entire life cycle. It is important to consider the full life cycle in this type of analysis as costs may be higher upfront for greater benefits later on.

Full Cost Accounting

Full cost accounting allows for the implementation of environmental, health, and social assets to be considered in the economic costs and benefits of decision making. It allows for complete end-to-end cost analysis of producing products or services. The goal is not necessarily to monetize their value, but rather to better understand their impact on an ecosystem and society, and potential ability to create value. Full cost accounting can help in sustainability by identifying areas where reducing environment impact could create value and confer monetary and non-monetary benefits that can be translated into economic terms. It allows for better management of the natural and social resources in our world today and a way to achieve more sustainable outcomes.

14

Fc

Full Cost Accounting

Harm Charge / Carbon Tax

32

Hc

Harm Charge / Carbon Tax

Carbon tax is a tax placed on the burning of fossil fuels, or carbon-based fuels, and corresponds to greenhouse gas emissions. The fee is either placed on the producers or passed along to the consumers for the carbon emissions produced from the burning of fuel. It can be implemented in a number of different facets such as through costs associated with home heating, flights, and shipping capacities. A harm charge is a similar concept only less specific. It places a charge on any practice harmful to the environment or human health such as a chemical spill, using non-sustainable products, or water pollution. The charges or taxes are utilized in order to serve as a disincentive to practices harmful to the climate, the environment, and human health, helping to shift thinking toward more sustainable practices. Further, the revenue generated from these charges can be used to invest in sustainability practices elsewhere in the system.

Depletion Charge

Similar to harm charge, depletion charge presents an opportunity to internalize an externality. In this case, the externality is the consumption of finite resources. Depletion charge presents an opportunity to incrementally increase the economic cost associated with the use of a finite resource by factoring in scarcity. That is, using the next amount of a finite resource would be increasingly expensive to disincentivize its ongoing use. This incremental charge does not need to be linear, and could itself increase with ongoing use of the finite resource, rapidly leading to a price that is cost prohibitive, rendering that finite resource economically infeasible.

50

Dc

Depletion
Charge

68 Rf
Sustained Research Funding

Sustained Research Funding

Scientific investigations, studies, and breakthroughs will be necessary to move away from the unsustainable trajectory that we are on. These investigations, in many cases, may require sustained efforts either due to the difficulty of the challenge or the inherent nature of longitudinal insights. These research efforts are particularly fragile to disruptions caused by the unpredictable modalities of research funding schemes whether in the public or private sector. The imperative of sustained, predictable funding support will be an essential element of achieving the kinds of insights and inventions necessary for sustainable prosperity.

Capital Investment

The chemicals and material manufacturing infrastructure that makes everything from building materials to wind turbines is very capital intensive. Current schemes for capital investment (e.g., venture, private equity), have the common goal of achieving targeted returns while minimizing risk. This model significantly favors incrementalism which makes it far easier to understand and analyze the risk profile for technological investments. It also favors low capital-intensive projects such as software and "app" projects and disfavors projects such as large-scale infrastructure and manufacturing. Achieving sustainability goals necessitates transformative technologies and dramatically reconstituted sustainable infrastructure at unprecedented scale requiring wise, patient capital at a significantly increased level.

86

Ci

Capital Investment

METRICS

For metrics to be useful in sustainability, they must have attributes that reflect sustainability itself. They must be both quantitative and qualitative. Looking at the historical metrics, and the way that professions often "worship at the altar" of efficiency metrics, it often seems that society has outsourced our decision making to quantifiable reductionist metrics. When thinking about achieving sustainability goals, we need to far exceed those simplistic metrics, and couple them with a more qualitative, and often, uncomfortable new generation of metrics. These metrics will move beyond mere quantities and instead provide insights into the character and inherent nature of our energy sources, of our stored energy storage systems, of our materials, and our basic chemicals. Sustainable metrics must be applicable and relevant across time and space, systems-based, and able to combine both reductionism and integrative systems thinking.

7
Ae
Atom Economy

15
Ef
E-Factor

33
Ff
F-Factor

51
Ql
Qualitative Metrics

69
Qn
Quantitative Metrics

87
Bb
Chemical Body Burden

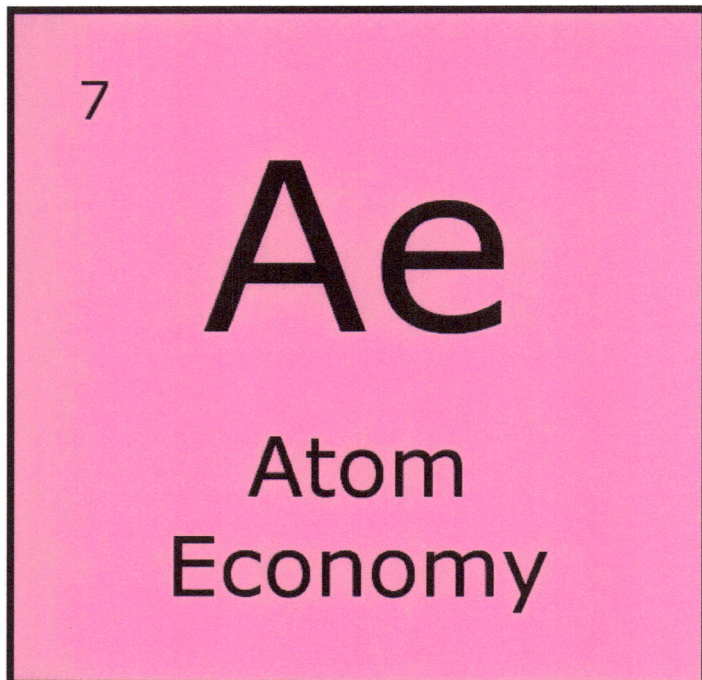

Atom Economy

Measuring the efficiency of chemical reactions is necessary to compare alternative synthetic routes to products in terms of environmental and economic costs. Percentage yield has historically served this function, as it compares the (predicted) theoretical and actual product quantity. However, a high yield is not sufficient to identify environmentally-preferable synthetic routes because there can be significant waste generated in a reaction that produces close to the predicted amount of product. As such, we need an additional metric, atom economy, to measure how much of the reactant atoms actually form the final product whereby:

$$\% \text{ Atom Economy} = \frac{\text{Molar Mass of Product}}{\text{Molar Mass of All Reactants}} \times 100\%$$

This is effectively guiding a chemist to pursue pollution prevention at the molecular scale. The higher the atom economy, the lower the amount of waste product formed. Both the yield and the atom economy should be taken into account when designing a green chemical process.

E-Factor

The amount of waste produced in the manufacture of a chemical product or a product of any kind should be minimized. While one would intuit that this measure was always part of manufacturing efficiency, it was not. The introduction of the calculation of the E-Factor took the original form of:

E-Factor = kg waste / kg product

and can be adjusted to include a variety of aspects of the manufacturing process. E-factor provides information that was historically neglected but is critical to moving toward more efficient and effective chemical production.

15

Ef

E-Factor

F-Factor

No one ever bought a chemical. They bought function or performance. They wanted the service the chemical provided. With that realization, a metric, F-Factor, has been developed to recognize and quantify the desire to achieve maximum function with the minimal amount of chemical used.

This approach has been expanded in the discussions of the chemical equivalent of Moore's Law.

The desire is for the value of F to be as large as possible by increasing the functional performance and/or decreasing the amount used to achieve some functional performance. This drives the system toward ideality where you get all of the function of a chemical or product without the existence of a chemical or a product; the chemical analogy of getting the function of the telephone without the need for telephone wires on every street.

$$F = \frac{\text{Function}}{\text{kg of chemical}}$$

33
Ff
F-Factor

Qualitative Metrics

While most of traditional assessment is based around quantitative metrics, qualitative metrics may be equally necessary in providing understanding related to sustainability. The nature and the character of aspects of sustainable chemistry are not always reductionist exercises. The renewability of a feedstock, the toxicity of a molecule, the environmental justice implications of siting a factory, market acceptance of an energy technology etc., may all have qualitative aspects that are critically important. While qualitative metrics may be less rigorous and involve integrative systems thinking that is outside traditional analytical frameworks, it is possible that they are more closely linked to the interconnected nature of sustainability systems and goals as outlined in the United Nations Sustainable Development Goals.

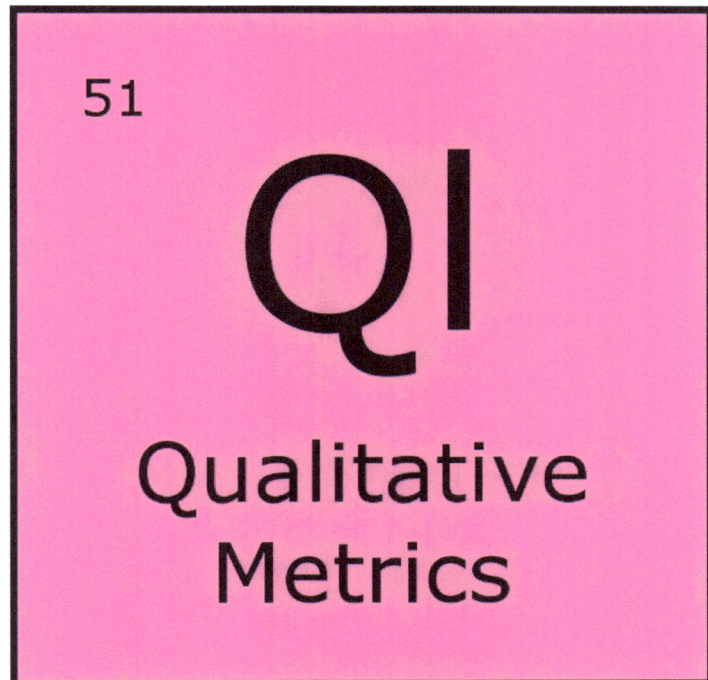

51

QI
Qualitative Metrics

Quantitative Metrics

69

Qn

Quantitative Metrics

Quantitative Metrics

One of the most active areas in analytical tools related to green chemistry and engineering is that of quantifiable metrics. There are important measures of (process) mass intensity ((P)MI), reaction mass efficiency (RME), carbon efficiency (CE), innovative green aspiration level (iGAL), and others. These quantifiable metrics can be useful and informative in answering specific questions of efficiency and environmental impact when a reductionist analysis is necessary to inform improvement. It is as important to know what data or information the quantifiable metrics are providing as it is to understand what questions they aren't addressing. The strengths of quantifiable metrics are essential and their weaknesses must be equally respected.

Chemical Body Burden

Chemical Body Burden refers to the measurement or load of chemicals in the body. This load can be detected through blood, urine, breastmilk sampling or any number of biomonitoring activities. Chemicals in the body could have numerous harmful effects and could weaken the immune system, making the body more susceptible to disease. The burden may be due to bioaccumulation through various mechanisms of exposure. Understanding the amount and types of chemicals in the body can help us know what chemicals are mobile in the environment. Chemical Body Burden is important to sustainability because it can make known the chemicals with the greatest bioaccumulation in nature and greatest exposures to humans, helping to inform future design of chemicals and products that avoid these unintentional impacts.

87

Bb

Chemical Body Burden

POLICIES AND REGULATIONS

The landscape of technologies in the chemical enterprise is not merely shaped by scientific and engineering solutions alone but rather in combination with the environment of regulation, policies, and laws that construct the social context in which they operate.

Policies and regulations can accelerate or retard sustainability solutions and they can protect or help displace entrenched unsustainable technologies.

The development of regulations and policy that will empower and enable green chemistry and green engineering to succeed in society will be necessary to remove the obstacles and inertia that keep the status quo in place.

| 8 **Pr** Extended Producer Responsibility |
| 16 **Pb** Property Based Regulation |
| 34 **Ct** Chemical Transparency |
| 52 **Cl** Chemical Leasing |
| 70 **Se** Self-Enforcing Regulations |
| 88 **I** Innovation Ecosystem Translation from Lab to Commerce |

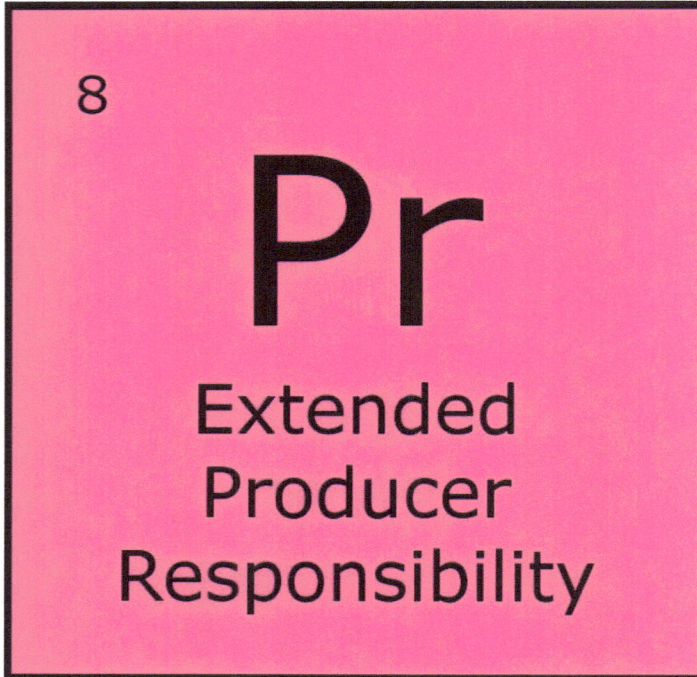

Extended Producer Responsibility

Extended Producer Responsibility refers to the responsibility a producer holds to design their product in a way that reduces negative environmental and health impact. It also places end-of-life management on the producer and not the consumer. This is especially important to sustainability. If the producers, who are responsible for what the environment and the population is exposed to, are not held responsible for ensuring their products are healthy and environmentally friendly, it becomes a problem of remedy rather than stopping it at the source. Extended producer responsibility helps to manage and mitigate end-of-life waste and all pollution generated from products at all stages of its life cycle.

Property-Based Regulation

The history of chemical regulation involved constructing lists of chemicals that were considered too hazardous or too risky and imposing some types of controls on them. Chemical-by-chemical regulation is slow, costly, inefficient, and inadequate. The nature of the concern for a chemical is not based on its chemical name but rather on its combination of properties. Some combinations of properties may lead a chemical to be bioavailable, others to be persistent, still others to be reactive or explosive. Since it is the combination of these properties that cause the concern, it is the properties that should be the basis of regulation. In this way, the chemicals of concern can be addressed proactively while providing certainty to the regulated community and critical guidance to the molecular designers of future products.

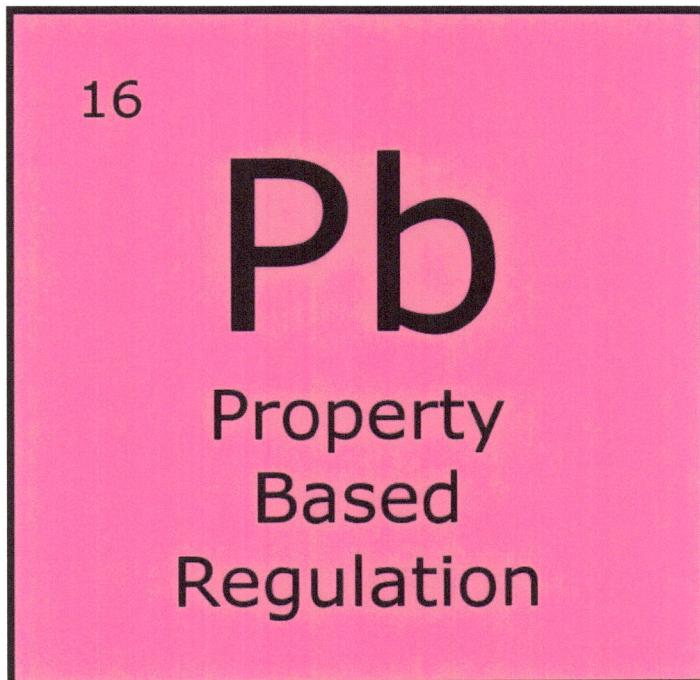

16

Pb

Property Based Regulation

Chemical Transparency

Chemical transparency refers to the disclosure of all chemicals and ingredients in all products. This is becoming exceedingly important among consumers. Consumers want to know what is in their chemical products and as part of their decision to trust the brands they use. This pressure from consumers and government can help to hold manufacturers accountable and aid in the development of safer, more benign chemicals, and continually move toward developing more sustainable products. Not only does this give consumers the option to choose safer, more environmentally-friendly products, but it also allows for easier and better exchange of practices among different companies. There are significant efforts underway to standardize the reporting of chemical ingredients in a variety of products to aid in data collection and product comparison.

Chemical Leasing

There is very little value to the ownership of the vast majority of chemicals unless they are used for their intended purpose. And yet, because of the traditional sell/own business model, the waste that results from over-purchasing is significant and systemic. Further, the system is constantly driving toward selling more chemicals to make more profit, driving toward designed obsolescence and short-term use. The chemical leasing model is one where the "sellers" supply the service or function rather than selling a chemical. In this model, instead of selling as much chemical as possible (resulting in excess and waste), the motivation is to use as little of a chemical to accomplish the desired service as the same profit can be generated from delivering much less product. Chemical leasing is a business model whose effectiveness has been demonstrated at large scale and has extensive potential to be expanded.

| 52 |
| **Cl** |
| Chemical Leasing |

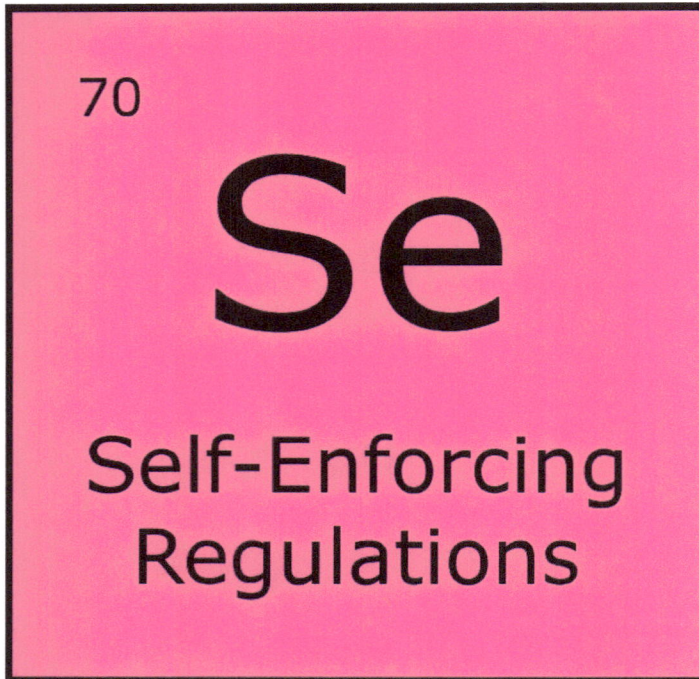

Self-Enforcing Regulations

70

Se

Self-Enforcing Regulations

The idea that regulations and laws to protect the environment and human health can be effectively enforced by government inspectors and officials has been belied by experience. This historical model has demonstrated over the past half-century that most of the damage is not detected and when it is detected it is at a timeframe that is inadequate to prevent the often tragic consequences. With the advent of sensors, real-time in-process controls, big-data analysis, and machine-learning, it is now possible to rethink models for enforcement. Instead of sampling hazardous waste sites for laboratory analysis to determine levels of contamination, there are now possibilities for integrated, networked monitoring systems. New processes can be designed such that they only can function as long as emissions are below certain levels for various contaminants. The emerging field of self-enforcing regulations can also build in predictability and reduced economic and time burden for the regulated community while ensuring enhanced effectiveness in achieving the goals of the regulation.

Innovation Ecosystem - Translation from Lab to Commerce

In order for a discovery or innovation to have an impact on the world, it virtually always needs to be enabled by an innovation ecosystem that incorporates the essential enabling elements. These include support for basic research and development and that R&D needs to be commercialized through thoughtful and risk-taking investment. These roles can be filled by various actors ranging from public sector government agencies to private sector investors. Appropriate and just intellectual property considerations need to be supported by governance and returns on investment need to be realized through adequate financial systems. With key roles and responsibilities assured, the innovation ecosystem can be used to bring about the transformative innovations in the chemical enterprise that are needed to advance sustainability.

88

I

Innovation Ecosystem Translation from Lab to Commerce

TOOLS

Increasing the development of tools to enable Green Chemistry and Green Engineering has expedited the adoption of more sustainable choices in the selection and innovation of products and processes. These resources allow chemists, engineers, and other professionals to pursue the incorporation of green chemistry into a wider range of products and processes with greater confidence in their decision-making and more credibility in building the business case for implementing these innovations.

9	**Ea** Epidemiological Analysis and Ecosystem Health
17	**Aa** Alternatives Assessment
35	**Lc** Life Cycle Assessment
53	**So** Solvent Selection Screens
71	**Cf** Chemical Footprinting
89	**Et** Education in Toxicology and Systems Thinking

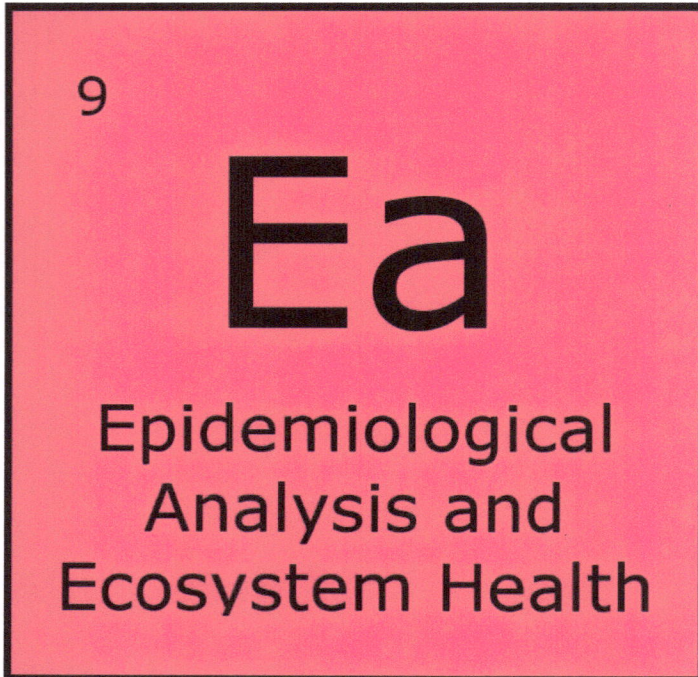

Epidemiological Analysis and Ecosystem Health

Epidemiological Analysis refers to the understanding and analysis of how different populations are affected by various risk-factors, disease, and other health outcomes. Ecosystem Health refers to the health or condition of the environment and all living organisms in that environment. That is, its susceptibility/resiliency to natural disaster and ability to sustain life defined by a number of indicators unique to the ecosystem. In conjunction, these two concepts define the health of an area and all the living organisms within it. These concepts are key to sustainability because they define the successfulness of sustainability practices. If something is being carried out in an unsustainable way it will be directly reflected in the health of the population and environment.

9

Ea

Epidemiological Analysis and Ecosystem Health

Alternatives Assessment

Alternatives Assessment is a technique aimed at minimizing harm by assessing all options and solutions and understanding the consequences of each. It helps to characterize hazard based on health and environmental information. It is often utilized in risk assessment and as a decision-making approach. One common application is a chemical alternatives assessment in order to aid in choosing the safer chemical over a more hazardous one with the aim of avoiding regrettable substitution.

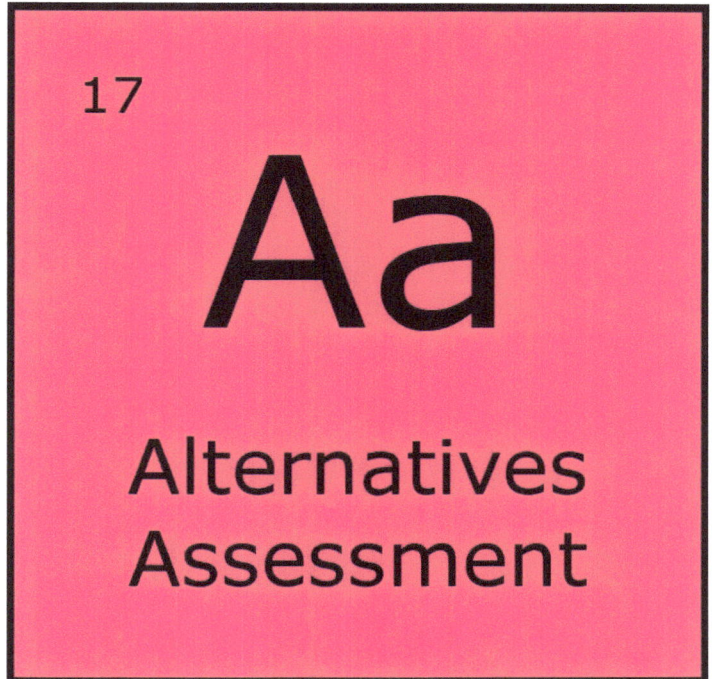

17

Aa

Alternatives
Assessment

Life Cycle Assessment

35

Lc

Life Cycle Assessment

Life Cycle Assessment determines the total environmental impact of a product from beginning to end of life, or cradle-to-grave. This is done through accounting for all material and energy-related inputs and outputs throughout the life of a product. It helps to measure all steps of production and use, and their subsequent environmental and health implications. Life Cycle Assessment provides a producer information for all direct and indirect environmental impacts associated with their products and processes, illuminating areas for improved design choices or making decisions between one chemical product or process and another.

Solvent Selection Screens

Solvents are often critical to a chemical process as a medium in which to facilitate contact of reactants and manage energy flows. They are also often the deciding aspect in the cost and environmental impact of a chemical process. However, when it comes to choosing the right solvent there are often many options, all which have different properties and different environmental impacts. A solvent selection tool identifies different solvent options using a variety of statistical, regression, and structure/property-based strategies, and then provides comparisons through various graphical outputs and shortlists. It allows the user to apply filters that select for certain properties and pick the best possible solvent for the situation. It also can aid in sustainability through improving industrial process and aiding in the evaluation of solvents in regard to their environmental and health impact.

53

So

Solvent
Selection
Screens

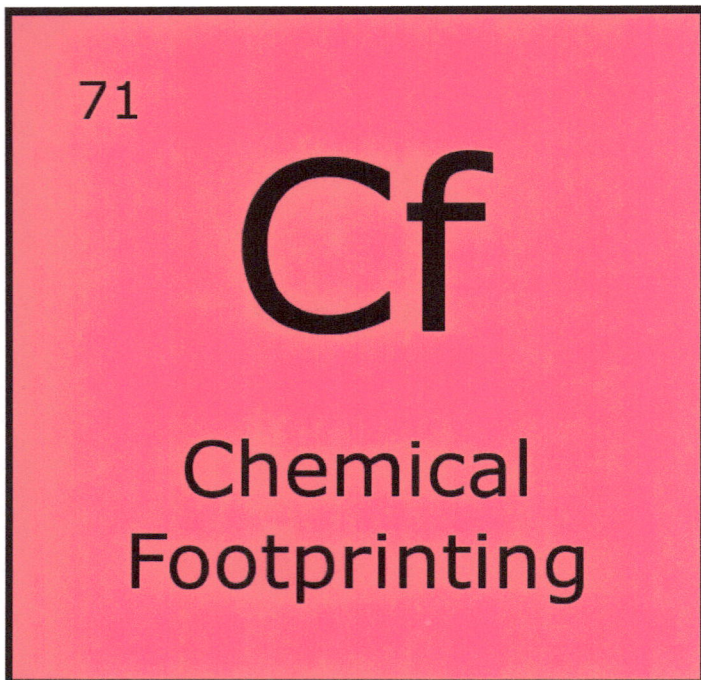

Chemical Footprinting

71

Cf

Chemical Footprinting

Chemical Footprinting allows for comprehensive, full life cycle management of chemicals being used by companies today. It gives companies the tools and ability to better understand what chemicals are in their products and a metric to better manage the safety and environmental impact of those chemicals. This benchmarking tool is important in sustainability practices because it helps to create a common standard for all to follow, while also reducing chemical risk, identifying areas for improvement, and measuring the progress being made.

Education in Toxicology and Systems Thinking

In most professions, the people that create something have responsibility for the safety of what they create. Th is is not true in the fi eld of chemistry. All chemists need to understand the consequences to the world and its inhabitants of what they create and study. Yet, virtually no training programs for chemists include the requirement of a course in molecular toxicology.

Th e basic principles of chemical dose-response, bioavailability, bioaccumulation, and biomagnifi cation need to be understood by chemists as well as the molecular level understanding of the fundamentals of acute and chronic toxicity, carcinogenicity, and endocrine disruption, among all relevant human health endpoints.

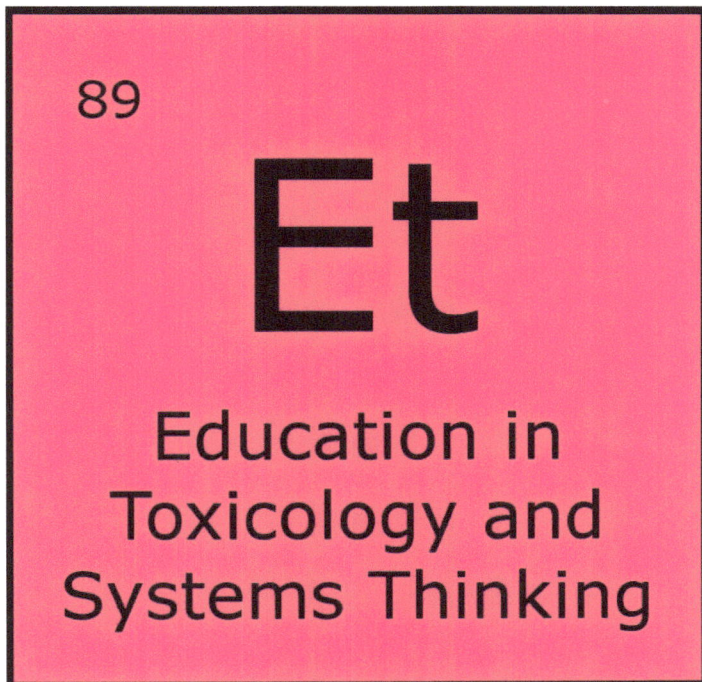

89

Et

Education in Toxicology and Systems Thinking

Noble Elements

Molecular scientists have been given the gift of the ability, the knowledge, the capability, the awareness, and the intellect to manipulate matter, to transform it from one thing to another, to give it different properties. Most of that gift does not come from us, it's not due to us. It is through birth, through opportunity, through circumstance and yes, a little hard work that we are given the gift of insight of chemistry. With that gift comes responsibility; we no longer have the option to say, "I don't care." We no longer have the option to say, "I don't know." We no longer have the option to say, "This is not relevant to me." Because we do and it is. And with that level of awareness, we have that responsibility. And these are our noble, not Nobel, our noble goals that we need to strive toward. It is with this knowledge that we must act.

NOBLE GOALS

There are some concepts that are transcendent. These considerations rise above the immediate economic or political concerns of the day. Noble elements are those that are grounded in moral imperatives that are shared across cultures and across time. These elements find their basis in values, justice, and trans-generational equity. Chemistry can impact all of these issues either positively or negatively and therefore these goals must consciously enter into our decision-making and our designs.

2
Ho
Hippocratic Oath for Chemistry

10
P
Design for Posterity

18
Lp
Life-Compatible Products & Processes

36
Z
Zero Waste

54
Fi
Chemistry is Equitable and Fully Inclusive

72
De
Benefits Distributed Equitably

90
K
Extraordinary Chemical Knowledge Comes with Extraordinary Responsibility

Hippocratic Oath for Chemistry

First, do no harm. The products that are made and the processes that make them and the resources that they come from will do no harm to the planet, both the biosphere that occupies it and the geosphere that sustains all living things. There will be no harm to the workers obtaining the feedstocks and transforming them, to the consumers using them, to the communities and populations, human and non-human alike, that may be exposed to them. There will be as much thought and care put into the consequences of chemistry as there is into the design and invention of the chemistry.

Design for Posterity

We borrow this planet and everything that is in it from the future. We own nothing. We owe everything. Our debt is repaid through the care exhibited by thoughtful use and design. Perpetuating flawed unsustainable systems based on limited knowledge or short-sighted perspectives is design for the past which simply enshrines ignorance and errors.

Constant change to reflect the highest levels of awareness that can lead to a more sustainable world is the path toward being respected by the future and ensuring that there will be a future to be respected by.

10

P

Design for Posterity

Lp

18

Lp

Life-Compatible Products & Processes

Life-Compatible Products & Processes

Nature is conducive to life. We are part of Nature and must assume our role as being conducive to life. The chemistry we discover and the chemicals we introduce into the world must reflect this role. The idea of legally acceptable toxicity and poisoning, socially acceptable degradation of ecosystems, and tolerable rates of species extinctions are frameworks that are flawed, illogical, and incompatible with our role in Nature.

Zero Waste

Waste is a man-made concept. In Nature outside of man, there is no waste. Evolutionary brilliance ensures that the waste from one organism will be utilized at high value to another organism. Waste in our chemical and material world is simply a material or energy for which a valuable use has not been discovered or implemented. While we will never defeat thermodynamics in attaining perfect cycles and entropy will always win, the cycles and systems that can be constructed can strive to continuously move toward the perfect goal of zero waste.

36

Z

Zero Waste

Chemistry is Equitable and Fully Inclusive

54

Fi

Chemistry is Equitable and Fully Inclusive

The natural laws of chemistry have no predisposition to any gender, race, creed, religion, nationality, ability, nor orientation. Yet the participation within the chemical enterprise has historically been dominated by a small demographic sliver of the population. The science and application of chemistry can only benefit by a maximum breadth of perspectives, skills, experiences, cognitive approaches, and values of a diverse community with a culture of inclusion. No endeavor, scientific or otherwise, can be 100% effective if it excludes so much of its talent. It would be as absurd to limit participants of a scientific field to individuals of a certain height or weight as it would be to limit them by their gender or ethnic background.

Full inclusion of all groups in the chemical enterprise should be pursued as a pathway to genuine excellence and not merely the narrow historical definitions of excellence as posited by those who have historically dominated the field.

Benefits Distributed Equitably

The benefits of chemistry are immense and have revolutionized the quality and length of human life. But not for everyone. The benefits of chemistry and chemicals are not distributed equally. Large percentages of the population have borne the burdens of a chemical-intensive society where the smallest percentage of the population has received benefits with virtually no burden.

Through thoughtful green and sustainable chemistry, new products will be agile, accessible, and adaptable and thus able to benefit people in the widest possible range of demographics, geographies, socio-economic, and logistical circumstances.

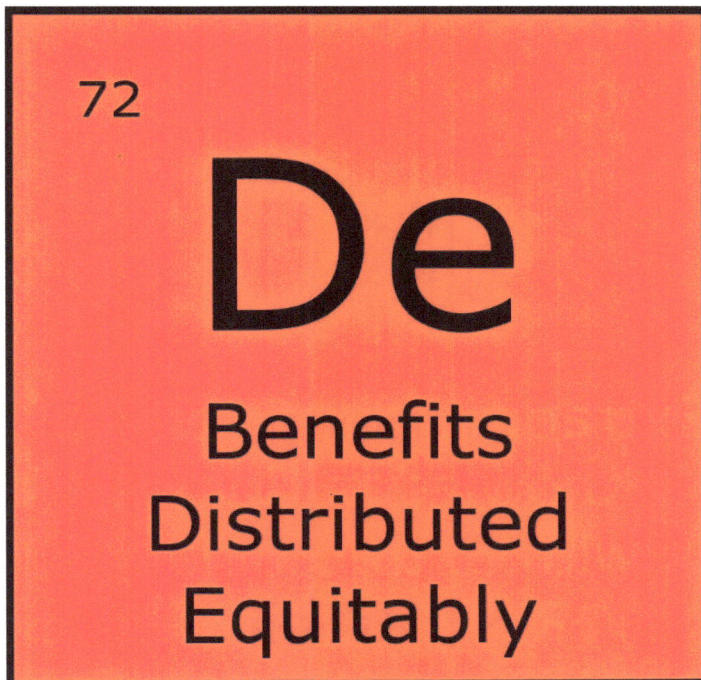

72

De

Benefits Distributed Equitably

90

K

Extraordinary Chemical Knowledge Comes with Extraordinary Responsibility

Extraordinary Chemical Knowledge Comes with Extraordinary Responsibility

A diminishingly small percentage of the population has the knowledge and understanding of how to manipulate matter at the molecular level. Those possessing this knowledge came to it through a confluence of the gift of sufficient intellectual capacity and the good fortune of having access to some type of educational framework not of their making. The power of chemistry is daunting and world-changing. It impacts societies, oceans, and atmospheres.

With these two gifts comes a responsibility to use the power of molecular manipulation for good and not ill. To build and not destroy. To heal and not to harm.

Valency

Every chemist knows about "valency." Valency – electronic structures that make the forming of bonds possible — is critical to making molecules. But valency has another definition: "the capacity of one person or thing to react with or affect another in some special way as attraction or the facilitation of a function or an activity." As we reflect on the elements of the Periodic Table of Green and Sustainable Chemistry, how can we make the elements react? Professional communities and social structures are often oriented around individual elements, from catalytic chemists to policy makers. In what ways are we going to facilitate the reaction with other elements? In what ways are we going to form these new bonds? In chemistry, covalency and covalent bonding is the essence of sharing; a sharing of each to form a better, stronger whole. What are these new molecules that will be formed? What are the new molecules that are being formed by these bonds? Education in Toxicology and Systems Thinking

In most professions, the people that create something have responsibility for the safety of what they create. This is not true in the field of chemistry. All chemists need to understand the consequences to the world and its inhabitants of what they create and study. Yet, virtually no training programs for chemists include the requirement of a course in molecular toxicology. The basic principles of chemical dose-response, bioavailability, bioaccumulation, and biomagnification need to be understood by chemists as well as the molecular level understanding of the fundamentals of acute and chronic toxicity, carcinogenicity, and endocrine disruption, among all relevant human health endpoints.

CONCLUSION

The Periodic Table of the Elements of Sustainable and Green Chemistry is not complete, and perhaps never will be. Just as Mendeleev's table had many gaps when it was introduced in 1869, this table, too, has unknown elements yet to be discovered. Every element is and will be critically important as we strive to realize the noble goals.

To attain the full power and potential of chemistry to improve the world, we need to begin with the fundamental science of green chemistry and green engineering and implement all of the tools, frameworks and perspectives contained in the elements of sustainable chemistry. You cannot achieve sustainable chemistry without green chemistry. Green chemistry will not be implemented at scale without the other elements of sustainable chemistry. Like all human endeavors and actions, all of these efforts in chemistry need to take place within an ethical, humanitarian, and moral framework. Just like the countless possible substances of the known universe are comprised of the known elements of the Periodic Table, there are countless possible paths to a sustainable future when employing the elements of green and sustainable chemistry.

BIBLIOGRAPHY

Anastas, P. T. and Warner, J. C.; Green Chemistry: Theory and Practice, Oxford University Press, 1998.

Anastas, P. T. and Zimmerman, J. B.; Design through the 12 principles of green engineering, Environ. Sci. Technol., 2003, 37, 94A–101A

Anastas, P. T. and Zimmerman, J. B.; The periodic table of the elements of green and sustainable chemistry. Green Chemistry. 2019;21(24):6545-66.

Benyus, J. M.; Biomimicry: Innovation inspired by nature, Morrow, New York, 1997.

United States Department of Health and Human Services, Centers for Disease Control and Prevention; Fourth Report on Human Exposure to Environmental Chemicals, 2009.

MacArthur Foundation; Towards the Circular Economy volumes 1-3, 2012-2014.

National Research Council; A framework to guide selection of chemical alternatives. National Academies Press, 2014.

Poliakoff, M., Licence, P. and George, M. W.; A New Approach to Sustainability: A Moore's Law for Chemistry, Angew. Chem., Int. Ed., 2018, 57, 12590–12591.

Sheldon, R. A.; The E factor 25 years on: the rise of green chemistry and sustainability, Green Chem., 2017, 19, 18–43

Trost, B. M.; The atom economy–a search for synthetic efficiency, Science, 1991, 254, 1471–1477

United National International Development Organization; Chemical Leasing: Global Promotion and Implementation of Chemical Leasing Business Models in Industry, Vienna, 2016.

United Nations General Assembly; Transforming our world : the 2030 Agenda for Sustainable Development, A/RES/70/1, 2015.

Index

T

V

W

Z

About the Authors

Paul T. Ansatas, PhD

Paul Anastas is known widely as the "Father of Green Chemistry" because he coined and defined the term "green chemistry" in 1991 and launched the first green chemistry research funding programs while at the United Stated Environmental Protection Agency. A synthetic organic chemist by training, Dr. Anastas served in the Administration of four U.S. Presidents and most recently was appointed by President Barack Obama to be the Chief Scientist and head of research and development for the U.S. Environmental Protection Agency. He has been a professor at Yale University since 2007 and is the founding Director of the Center for Green Chemistry and Green Engineering at Yale.

Jule B. Zimmerman, PhD

Julie Zimmerman is an internationally recognized engineer whose work is focused on advancing innovations in sustainable technologies. She holds joint appointments as a Professor in the Department of Chemical and Environmental Engineering and School of the Environment at Yale University. She also serves at the Senior Associate Dean for Academic Affairs at the Environment School. Julie serves as the Editor in Chief for Environmental Science & Technology. Previously, she served at the U.S. Environmental Protection Agency creating the national sustainable design competition, P3 (People, Prosperity, and Planet) Award, which has engaged design teams from hundreds of universities across the United States.